Liquid Crystal on Silicon Devices

Liquid Crystal on Silicon Devices: Modeling and Advanced Spatial Light Modulation Applications

Special Issue Editors

Andrés Márquez
Ángel Lizana

MDPI • Basel • Beijing • Wuhan • Barcelona • Belgrade

MDPI

Special Issue Editors

Andrés Márquez
University of Alicante
Spain

Ángel Lizana
Universitat Autònoma de Barcelona
Spain

Editorial Office
MDPI
St. Alban-Anlage 66
4052 Basel, Switzerland

This is a reprint of articles from the Special Issue published online in the open access journal *Applied Sciences* (ISSN 2076-3417) from 2018 to 2019 (available at: https://www.mdpi.com/journal/applsci/special_issues/LCoS).

For citation purposes, cite each article independently as indicated on the article page online and as indicated below:

LastName, A.A.; LastName, B.B.; LastName, C.C. Article Title. *Journal Name* **Year**, *Article Number*, Page Range.

ISBN 978-3-03921-828-8 (Pbk)
ISBN 978-3-03921-829-5 (PDF)

Contents

About the Special Issue Editors

Andrés Márquez received his MSc (1997) and PhD (2001) degrees in physics from the Universidad Autónoma de Barcelona. Since 2000 he has been with the Group of Holography and Optical Processing, Universidad de Alicante (UA). He is Professor of Applied Physics in the Department of Physics, Systems Engineering, and Signal Theory (FISTS), where he has been Head of the Department. He is currently the Director of the Research Institute of Physics Applied to Sciences and Technologies at the UA. His research focuses on holographic recording materials and holographic memories, spatial light modulators based on liquid crystal, optical image processing, and diffractive optics. He has published more than 180 articles, and has more than 220 communications in congresses (16 invited). He is co-author of four book chapters, co-editor of five SPIE Proceedings Volumes, and co-inventor of one patent. He is senior member of OSA and Fellow of SPIE.

Ángel Lizana completed his MSc degree in Physics and PhD in Physics at the Autonomous University of Barcelona (Spain) in 2006 and 2011, respectively. His research interests include liquid crystal displays and their application to diffractive optics, as well as the design and implementation of polarimeters, and their application to biophotonics. He was a postdoctoral scientist in the PICM laboratory of the École Polytechnique (France) in 2011–12 and in 2013–2014, where he has worked on the optimization and implementation of a polarimeter working in the infrared spectrum and on the analysis of polarimetric information for its application to the characterization of samples. Dr. Lizana has been an Associate Professor at the Autonomous University of Barcelona (UAB) since 2016. He is member of the SPIE and of Sociedad Española deÓptica (SEDOPTICA). He was appointed General Secretary of SEDOPTICA in 2019.

![applied sciences logo] *applied sciences*

MDPI

Editorial

Special Issue on Liquid Crystal on Silicon Devices: Modeling and Advanced Spatial Light Modulation Applications

Andrés Márquez [1,2,*] and Ángel Lizana [3,*]

1   Instituto Universitario de Física Aplicada a las Ciencias y las Tecnologías (IUFACyT),
    Universidad de Alicante, 03690 Alicante, Spain
2   Departamento de Física, Ingeniería de Sistemas y Teoría de la Señal, Universidad de Alicante, Ap. 99,
    03080 Alicante, Spain
3   Departamento de Física, Universitat Autònoma de Barcelona, 08193 Bellaterra, Spain
*   Correspondence: andres.marquez@ua.es (A.M.); angel.lizana@uab.es (Á.L.)

Received: 16 July 2019; Accepted: 26 July 2019; Published: 29 July 2019

1. Introduction

Since the first liquid crystal displays (LCDs) at the beginning of the seventies—based on the twisted-nematic cell configuration [1]—LC-based devices [2] have shown a great potential not only as a display technology, but also for spatial light modulation applications.

Among the different LC-based technologies, liquid crystal on silicon (LCoS) has become one of the most widespread technologies for spatial light modulation in optics and photonics' applications [3–6]. These reflective microdisplays are composed of a high-performance silicon complementary metal oxide semiconductor (CMOS) backplane, which controls the liquid crystal layer's light modulating properties . State-of-the-art LCoS microdisplays may exhibit a very small pixel pitch (below 4 µm), a very large number of pixels (resolutions larger than 4 K), and high fill factors (larger than 90%). They modulate illumination sources covering the UV, visible, and IR.

LCoS technologies are used as displays as well as polarization, amplitude, and phase-only spatial light modulators, where they achieve full phase modulation. Due to their excellent modulating properties and high degree of flexibility, they are found in all sorts of spatial light modulation applications, such as in LCoS-based display systems for augmented and virtual reality, head-up display, head-mounted display, projector, true holographic displays, digital holography, optical storage, adaptive optics, diffractive optical elements, super-resolution optical systems, optical metrology techniques, reconfigurable interconnects, beam-steering devices, wavelength selective switches in optical telecommunications, wave-front sensing of structured light beams, holographic optical traps, or quantum optical computing.

In order to fulfill the requirements in this extensive range of applications, specific models and characterization techniques are proposed. These devices may exhibit a number of degradation effects such as limited modulation range for high spatial frequency image content, interpixel cross-talk and fringing field, and time flicker, which may also depend on the analog or digital backplane of the corresponding LCoS device. Appropriate characterization and compensation techniques are then necessary.

2. Special Issue Papers

This special issue provides a collection of papers demonstrating the impact of LCoS microdisplays in current and future spatial light modulation applications. State-of-the-art in LCoS device technology, LC materials, modeling and characterization techniques are presented.

A complete review of LCoS origins, evolution, and applications is given by Chen et al. [7]. They cover both the interest in display applications and their developments as spatial light modulators (SLM), with special focus on phase-only spatial light modulation capabilities in digital holography or holographic video projection. The authors explain that the origin of LCoS backplane dates back to 1973, and the development of LCoS has been beneficial for full high definition displays and spatial light modulation. They start summarizing state-of-the-art developments of high-resolution panels, followed by addressing issues related to the driving frequency (i.e., liquid crystal response time and hardware interface), taking into account phase linearity control, phase precision, phase stability, and phase accuracy.

A good complement to the previous paper is given by Li and Cao [8]. The authors focus their review on the available characterization techniques to obtain the grayscale-phase response of LCoS devices. They demonstrate that precise calibrations are necessary since they greatly influence the results when applied in phase-only applications, including the majority of interesting cases such as holographic display, optical tweezers, lithography, etc. Due to limitations in the manufacturing process, the grayscale-phase response could be different for every single SLM device, even varying on sections of an SLM panel due to the screen or electrical addressing spatial inhomogeneities. They divide the numerous calibration methods into two categories: the interferometric phase calibration methods and the diffractive phase calibration methods. The main phase calibration methods are discussed and reviewed, comparing their possible advantages in different applications.

Regarding the analysis of LC materials enabled for LCoS devices, we consider the review paper by Andreev et al. [9]. They investigate ferroelectric liquid crystals (FLCs) of a new type developed for fast low-voltage displays and light modulators. They are helix-free FLCs, characterized by spatially periodic deformation of smectic layers and a small value of spontaneous polarization. Both theoretical models and experimental results are presented for modulation of light transmission, scattering, and phase delay with a high rate.

Alternative optically addressable spatial-light modulators (OASLM) are investigated by Pei et al. [10], which are very interesting for phase-only modulation applications. They focus on the influence of driving conditions on their phase-modulation ability. To this end, they use an equivalent circuit method and a system for measuring wave-front modulation that uses a phase-unwrapping data-processing method, and is constructed with a charge-coupled device and wave-front sensor. They demonstrate that wave-front on-line modulation with feedback control is possible with the OASLM and the corresponding monitoring system.

In previous years, there has been considerable interest in augmented reality (AR) displays. Huang et al. [11] review LCoS technology capabilities' new application in emerging AR displays. They start by reviewing the LCoS working principles of three commonly adopted LC modes. Then, the fringing field effect is analyzed, which is very important in very high resolution LCoS microdisplays. The novel pretilt angle patterning method for suppressing the effect is presented. They also show how to integrate the LCoS panel in an AR display system. Authors show that the application of LCoS in AR head-mounted displays and head-up displays is foreseeable.

In dealing with holographic projection and application of phase-only holography with SLMs for display applications, we consider the paper by Christmas and Collings [12]. They review the various approaches of producing dynamic holographic displays and show their particular proposal with a superior light efficiency and fault tolerance. Holographic displays favor small pixel devices, with LCoS devices as optimum candidates for these applications to become commercially feasible.

Davis et al. [13], in imaging applications, demonstrate a programmable zoom lens system where the magnification and sense of the image can be controlled without moving any parts. They use two programmable SLMs onto which they encode the required focal length lenses to achieve these results. They show both theoretical calculations and experimental results. The authors discuss the system's size limitations caused by the limited spatial resolution, and show how newer devices—LCoS with a very high resolution—would shrink the size of the system.

SLMs are also used in holographic data storage systems (HDSS). The paper by Martínez-Guardiola et al. [14] presents a method to characterize a complete optical HDSS. In this study, the authors identify the elements that limit the capacity to register and restore the information introduced by means of a LCoS microdisplay as the data pager. They further test whether the anamorphic and frequency-dependent effect is relevant in the application to HDSS, where nonperiodic binary elements are applied. They consider anamorphic patterns with different resolutions addressed to the LCoS. They show both the precharacterization results when no recording material is in the HDSS and when introducing a photopolymer as the recording material.

Another interesting application for LCoS devices is described by Rothe et al. [15]. Multimode fibers (MMF) are promising candidates for increasing the data rate while reducing the space required for optical fiber networks. The authors demonstrate a method for measuring the transmission matrix (TM) of a multimode fiber. It is based on mode-selective excitation of complex amplitudes performed with only one phase-only spatial light modulator. The light field propagating through the fiber is measured holographically and analyzed by a rapid decomposition method. The TM determines the amplitude and phase relationships of the modes, allowing us to understand the mode scrambling processes in the MMF, which can also be used for mode division multiplexing in telecommunications.

In the review paper by Zhang et al. [16], the authors demonstrate an LCoS self-calibration technique, from which they perform a complete LCoS characterization. They determine its phase–voltage curve by using the interference pattern generated by a digital two-sectorial split-lens configuration. They also determine the LCoS surface profile by using a self-addressed dynamic microlens array pattern. Once the LCoS is calibrated, they show both the application to microparticle manipulation through light optical traps created by a LCoS display. They also show the ability of the LCoS display to implement a holographic imaging system with a double-sideband filter configuration, so as to obtain dynamic holographic imaging of microparticles.

In the last paper of this collection, Pérez-Cabré and Millán [17] characterize a LCoS with a phase modulation much larger than 2π radians. Multiorder diffractive optical elements, displayed on the LCoS SLM with the appropriate phase modulation range, enable the design and experimental demonstration of an achromatic multiorder lens. They show that the residual chromatic aberration is reduced to one-third that of the chromatic aberration of a conventional first-order diffractive lens.

3. Perspectives

This Special Issue contains eleven papers addressing a wide variety of topics dealing with liquid crystal on silicon (LCoS) devices. The review and research papers provide a good insight into the present developments in this topic. We note that the availability of more mature technological devices, with ever decreasing pixel sizes and more stable operation, indicate that LCoS will widen their range of applicability even more in the coming years.

Funding: This research received no external funding.

Acknowledgments: The guest editors would like to thank all the authors and reviewers for contributing with their excellent papers and outstanding evaluation reports to this special issue. The guest editors also would like to thank the MDPI team involved in editing and managing this special issue. We would also like to thank Lucia Li, the contact editor for this special issue, for her very professional and generous support.

Conflicts of Interest: The authors declare no conflict of interest.

References

1. Schadt, M.; Helfrich, W. Voltage-dependent optical activity of a twisted nematic liquid crystal. *Appl. Phys. Lett.* **1971**, *18*, 127–128. [CrossRef]
2. Yeh, P.; Gu, C. *Optics of Liquid Crystal Displays*, 2nd ed.; Wiley: Hoboken, NJ, USA, 2009.
3. Collings, N.; Davey, T.; Christmas, J.; Chu, D.; Crossland, B. The Applications and Technology of Phase-Only Liquid Crystal on Silicon Devices. *J. Disp. Technol.* **2011**, *7*, 112–119. [CrossRef]

4.   Lazarev, G.; Hermerschmidt, A.; Kruger, S.; Osten, S. *LCoS Spatial Light Modulators: Trends and Applications, in Optical Imaging and Metrology: Advanced Technologies*; Osten, W., Reingand, N., Eds.; Wiley-VCH Verlag & Co.: Weinheim, Germany, 2012. [CrossRef]

5.   Zhang, Z.; You, Z.; Chu, D. Fundamentals of phase-only liquid crystal on silicon (LCOS) devices. *Light Sci. Appl.* **2014**, *3*, e213. [CrossRef]

6.   Lazarev, G.; Chen, P.-J.; Strauss, J.; Fontaine, N.; Forbes, A. Beyond the display: Phase-only liquid crystal on Silicon devices and their applications in photonics. *Opt. Express* **2019**, *27*, 16206–16249. [CrossRef] [PubMed]

7.   Chen, H.-M.P.; Yang, J.-P.; Yen, H.-T.; Hsu, Z.-N.; Huang, Y.; Wu, S.-T. Pursuing High Quality Phase-Only Liquid Crystal on Silicon (LCoS) Devices. *Appl. Sci.* **2018**, *8*, 2323. [CrossRef]

8.   Li, R.; Cao, L. Progress in Phase Calibration for Liquid Crystal Spatial Light Modulators. *Appl. Sci.* **2019**, *9*, 2012. [CrossRef]

9.   Andreev, A.; Andreeva, T.; Kompanets, I.; Zalyapin, N. Helix-Free Ferroelectric Liquid Crystals: Electro Optics and Possible Applications. *Appl. Sci.* **2018**, *8*, 2429. [CrossRef]

10.   Pei, L.; Huang, D.; Fan, W.; Cheng, H.; Li, X. Phase-Only Optically Addressable Spatial-Light Modulator and On-Line Phase-Modulation Detection System. *Appl. Sci.* **2018**, *8*, 1812. [CrossRef]

11.   Huang, Y.; Liao, E.; Chen, R.; Wu, S.-T. Liquid-Crystal-on-Silicon for Augmented Reality Displays. *Appl. Sci.* **2018**, *8*, 2366. [CrossRef]

12.   Christmas, J.; Collings, N. Displays Based on Dynamic Phase-Only Holography. *Appl. Sci.* **2018**, *8*, 685. [CrossRef]

13.   Davis, J.A.; Hall, T.I.; Moreno, I.; Sorger, J.P.; Cottrell, D.M. Programmable Zoom Lens System with Two Spatial Light Modulators: Limits Imposed by the Spatial Resolution. *Appl. Sci.* **2018**, *8*, 1006. [CrossRef]

14.   Martínez-Guardiola, F.J.; Márquez, A.; Calzado, E.M.; Bleda, S.; Gallego, S.; Pascual, I.; Beléndez, A. Anamorphic and Local Characterization of a Holographic Data Storage System with a Liquid-Crystal on Silicon Microdisplay as Data Pager. *Appl. Sci.* **2018**, *8*, 986. [CrossRef]

15.   Rothe, S.; Radner, H.; Koukourakis, N.; Czarske, J.W. Transmission Matrix Measurement of Multimode Optical Fibers by Mode-Selective Excitation Using One Spatial Light Modulator. *Appl. Sci.* **2019**, *9*, 195. [CrossRef]

16.   Zhang, H.; Lizana, A.; Van Eeckhout, A.; Turpin, A.; Ramirez, C.; Iemmi, C.; Campos, J. Microparticle Manipulation and Imaging through a Self-Calibrated Liquid Crystal on Silicon Display. *Appl. Sci.* **2018**, *8*, 2310. [CrossRef]

17.   Pérez-Cabré, E.; Millán, M.S. Liquid Crystal Spatial Light Modulator with Optimized Phase Modulation Ranges to Display Multiorder Diffractive Elements. *Appl. Sci.* **2019**, *9*, 2592. [CrossRef]

applied
sciences

MDPI

Review

Pursuing High Quality Phase-Only Liquid Crystal on Silicon (LCoS) Devices

Huang-Ming Philip Chen [1,*], Jhou-Pu Yang [1], Hao-Ting Yen [1], Zheng-Ning Hsu [1], Yuge Huang [2] and Shin-Tson Wu [2]

[1]   Department of Photonics, College of Electrical and Computer Engineering, National Chiao Tung University, Hsinchu 30010, Taiwan; yjp31167.di02g@nctu.edu.tw (J.-P.Y.); howard5132265@gmail.com (H.-T.Y.); zhn215222@gmai.com (Z.-N.H.)
[2]   College of Optics and Photonics, University of Central Florida, Orlando, FL 32816, USA; y.huang@Knights.ucf.edu (Y.H.); swu@creol.ucf.edu (S.-T.W.)
*   Correspondence: pchen@mail.nctu.edu.tw; Tel.: +886-3-513-1509

Received: 31 October 2018; Accepted: 14 November 2018; Published: 21 November 2018

Abstract: Fine pixel size and high-resolution liquid crystal on silicon (LCoS) backplanes have been developed by various companies and research groups since 1973. The development of LCoS is not only beneficial for full high definition displays but also to spatial light modulation. The high-quality and well-calibrated panels can project computer generated hologram (CGH) designs faithfully for phase-only holography, which can be widely utilized in 2D/3D holographic video projectors and components for optical telecommunications. As a result, we start by summarizing the current status of high-resolution panels, followed by addressing issues related to the driving frequency (i.e., liquid crystal response time and hardware interface). LCoS panel qualities were evaluated based on the following four characteristics: phase linearity control, phase precision, phase stability, and phase accuracy.

Keywords: liquid-crystal-on-silicon; spatial light modulator; holographic display; phase precision and stability; phase accuracy; spatially anamorphic phenomenon

---

1. Introduction

Photo-activated liquid crystal on zinc-sulfide (ZnS) [1], cadmium-sulfide (CdS) [2], and active-matrix addressed liquid-crystal-on-silicon (LCoS) [3] were first introduced by Hughes Research Laboratories for display applications in the early 1970s. Later, this approach was extended to the reflective spatial light modulator (SLM) [4,5]. The LCoS-based SLM panel resolution increased from 16×16 to 176×176 pixels, as was reported by the University of Edinburgh (Edinburgh, UK), in the period 1986–1989 [6]. In 1993, resolution was further increased to a 254×254 pixel array by DisplayTech (now CITIZEN FINEDEVICE (CFD) Co., LTD, Yamanashi, Japan). Their panel adopted the fast-switching ferroelectric liquid crystal (FLC) and implemented the field-sequential color system to generate a full color display. The ferroelectric-LCoS remains their distinctive product until today [7]. The projection display technology using LCoS as the core optical engine is not able to compete with digital micro-mirror device (DMD) technology with a resolution of 720p due to the cost structure. However, a high-resolution LCoS panel with a small pixel size can be commercialized owing to the advancement in integrated circuit (IC) technology during the past 10 years. Since 2000, LCoS has evolved from 720p to not only 2K1K Full High Definition (FHD) but also 4K2K resolution. Photos shown in Figure 1 of high-resolution display products are provided by Jasper Display Corp.

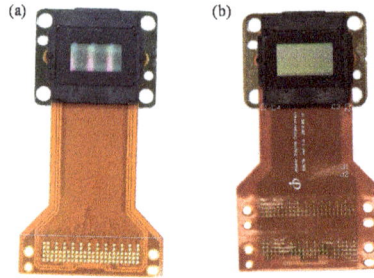

Figure 1. Photos of (**a**) 2K1K and (**b**) 4K2K panels from Jasper Display Corp.

Holographic near-eye displays using LCoS as a display panel opened a new era for virtual and augmented realities. Oculus Rift and HTC VIVE announced and demonstrated compelling 3D virtual reality (VR) experiences using JD5552 LCoS-SLM (Jasper Display Corp.; Hsinchu, Taiwan R.O.C.) in 2016 [8]. Microsoft demonstrated the HoloLens, its first self-contained augmented reality (AR) device, using PLUTO LCoS-SLM (Holoeye Photonics; AG, Berlin, Germany) in the same year [9]. In addition to Jasper Display Corp. and Holoeye Photonics, there are other worldwide providers of LCoS micro-displays, such as Hamamatsu Photonics, Meadowlark Optics, Santec Corp., and Himax. Each company has its own focus within its products. Jasper Display Corp. has its own digital drive 2K1K and 4K2K panels with corresponding drivers. Holoeye Photonics has a distinctive solution for 60 Hz 4K-SLM and computer-generated hologram (CGH) information input software development kit (SDK) interface design [10]. Hamamatsu provides pure-phase SLM solutions with high linearity and precision. It also provides SLMs for high-power laser applications [11]. Meadowlark Optics features the analog drive LCoS-SLM solution, which has a fast liquid crystal (LC) response time, a near-millisecond response, and a near-kHz picture frame rate [12]. Santec's LCoS-SLM phase modulator can provide ultra-high flatness (i.e., high phase precision) and a 10-bit linear curve [13]. Table 1 summarizes their unique products.

Table 1. Four main LCoS phase modulator providers and their unique products.

HOLOEYE Photonics AG	Hamamatsu Photonic	Meadowlark Optics	Santec Corp.
Products	Products	Products	Products
1. PLUTO-2: VIS-014, VIS-016, VIS-020, VIS-056 2. LETO: LETO-only 3. GAEA-2: VIS-036	1. X-10468: -01, -02, -03 2. X-13138:-01, -02, -03	1. P512: P-, PDM-, HSP-, HSPDM-, ODP-, ODPDM 2. P1920: P-, HSP-	1. SLM-100: -01 2. SLM-200: -01
Features	Features	Features	Features
1. UHD 4K-Panel (7000 ppi) 2. Driver Interface Up to 600 MHz(HDMI 2.0) 3. SDK Solution	1. Pure, Linear and Precise Phase Control 2. High-Power-Use	1. Analogy Driving Scheme 2. Fast LC Response Time 3. Sub-KHz PCIe Interface	1. Super Flatness Available: $\lambda/40$ 2. 10-bit LUT
* Only List for *VIS Light-Region-Use* LCoS-SLM ($T_{op.}$ = 35 °C) * BNS LCoS-SLM is acquired by Meadowlark Optics in 2014			

A phase-only spatial light modulator can be used as a key optical element for displays, adaptive optics for sensing, lithography, and telecommunication, as shown in Figure 2. The linearity of phase modulation, response time, phase precision, and phase stability are key characteristics for appraising or selecting phase-only LCoS-SLM panels for the applications designed. However, it is difficult to obtain a single LCoS possessing all the desired features for all applications. Their specification and performance optimization are application-driven. For example, response time is the current limitation in a holographic display. It may be more important than phase stability, because an excellent holographic image can still be achieved using a well linearly calibrated and high-phase precision LCoS-SLM panel. On the other hand, the phase stability of the LCOS-SLM is a vital property in the field of optical communication, because an instantaneous disturbance resulting from an unstable phase may cause misjudgment of the transmitted information [14]. Compared to fast response time, phase stability is much more important for optical communication applications.

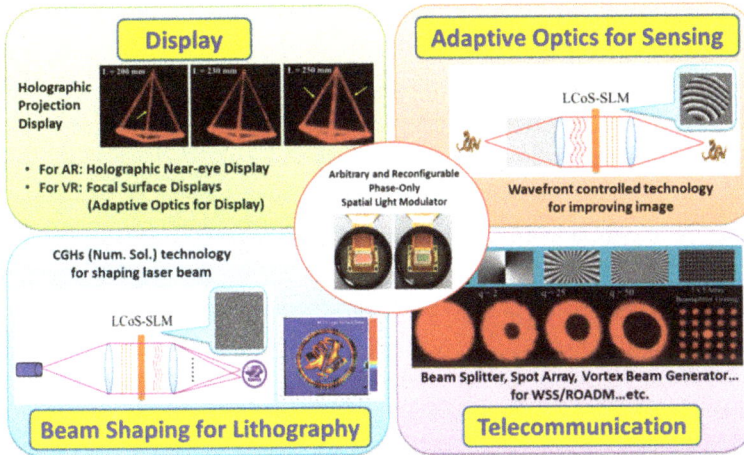

Figure 2. Schematic of various applications of phase-only liquid crystal on silicon spatial light modulator (LCoS-SLM).

Other applications may require precise phase modulation, such as processing of laser materials [15,16] or controlling the wave front for an image enhancement application [17]. We summarize the quantitative assessment indicators for the LCoS-SLM panels in Figure 3. The high-resolution panel is a trend for future phase modulation. The linearity of phase modulation is not only convenient for data acquisition, but also fulfills linear modulation in the diffraction equations derived from the scalar diffraction theory. The phase precision, phase stability, and phase accuracy can be further evaluated based on a linear phase calibrated panel. Real-time modulation depends on the LC response time, driver latency, and the calculation time of CGHs. In this review, we reveal current progress and approaches in designing and developing high-quality phase-only LCoS-SLM with the assessment of following criteria: panel resolution, driving frequency (LC response time and hardware interface), phase linearity control, phase precision, phase stability, and phase accuracy.

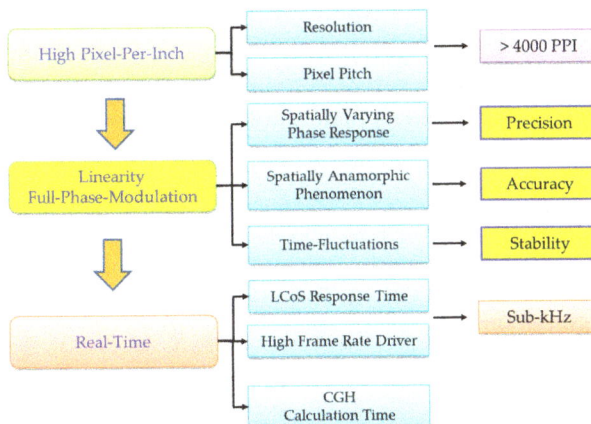

Figure 3. Quantitative assessment indicators for LCoS-SLM.

2. High Resolution and High Pixel-Per-Inch (PPI)

Achieving higher resolution with higher pixel density and smaller pixel pitch is a continuous goal of LCOS micro-displays, especially for near-eye VR display applications [18]. For holographic AR, higher diffraction, or larger image projection, a high-resolution panel is also required [19–21]. The small pixel size enables a large diffraction angle, and a large panel with high resolution empowers finer details of the projected image. The current resolution density in pixel per inch (ppi) and the pixel pitch of commercial LCoS-SLM panels are summarized in Figure 4. In particular, panels with high ppi (>5000) 4K2K or 8K4K LCoS possess a low input frame rate (≤60 Hz) or limited driving voltage (<3.3 V). In general, it is difficult to maintain a frame rate of 60 Hz in a 4K2K panel [22].

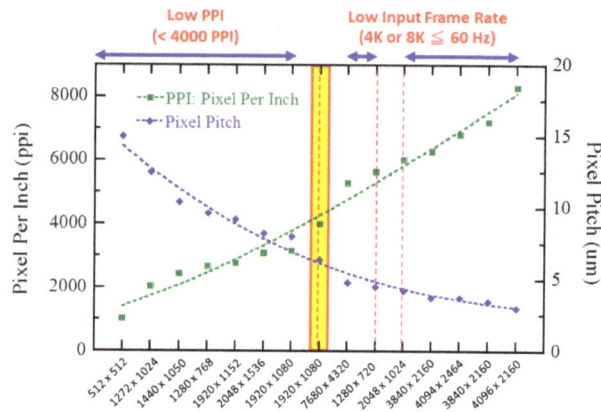

Figure 4. Pixel density and pixel pitch specifications for current LCoS-SLM integrated circuit (IC) backplanes.

Several approaches to achieve high frame rates in high-resolution panels (≥4000 ppi) have been realized recently. Yang et al. demonstrated 4000 ppi in a 0.55-inch 1920 × 1080 IC backplane with a pixel pitch of approximately 6.4 µm [23]. Abeeluck et al. demonstrated a digital drive LCoS micro-display with a pixel pitch of 3 µm and a high fill factor of 93.5% in resolutions from 1080p to 4K [24]. Such a small panel size can be easily integrated into an optical machine or an embedded system. The advantage of a panel with small pixels is that a larger diffraction angle can be obtained. In addition, two commercial products listed in Figure 4—the 0.26 inch 1280 × 720 resolution (Holoeye HED2200) and the 0.37 inch 2048 × 1024 resolution IC back panels (Himax Display; Tainan, Taiwan R.O.C.)—are suitable for panels with a large diffraction angle, while maintaining a high picture frame rate.

3. Switching Time for Phase Modulation

There are four major issues for holographic display applications: (1) the slow liquid crystal response time; (2) the small field of view (FoV); (3) the laser speckle effect degrading the image quality when using lasers as a backlight unit; and (4) a reduction in the frame rate for achieving field-sequential color displays [25]. There are many studies where the aforementioned issues have been addressed, such as: (1) 3–4 ms phase-only panel response time of thin cell gap at λ = 633 nm to reduce the dynamic reconstruction holographic image blur [26]; (2) the time-division multiplexing method to increase the FoV [27]; and (3) reduction of the speckle effect in image projection using temporal averaging techniques [28]. The ultimate turn-key solution to these challenges is the SLM panel with a sub-millisecond response time and a sub-kHz input frame rate.

Ferroelectric liquid crystal (FLC) modulators [29,30] and digital micro-mirror devices (DMD) [31] are two commercial spatial light modulators capable of achieving sub-millisecond response times. Significant quantization noise lowers the diffraction efficiency due to binary phase modulation. Several approaches have been proposed to achieve a fast response time, such as dual frequency LC (DFLC) [32], polymer network LC [33,34], and polymer-stabilized blue phase LC [35]. High driving voltage (greater than 10 V) and light scattering in the visible region are the two major drawbacks of these LC material systems. In addition, limited driving voltage in a high-resolution backplane often results in a slower LC response time. Lately, Wu's group reported that an average phase-to-phase response time of ~2 ms can be achieved using high Δn LCs in a thin cell gap of 1.7 μm at an operating voltage of 5 V at 40 °C [36,37]. The reported LCs can be used with 240 Hz frame rate without complicated overdrive or undershoot circuitries. Additionally, the effective birefringence of liquid crystals slightly decreases when the cell gap is below 1.8 μm due to the strong surface boundary effect of the alignment layers [38]. Thus, a higher birefringence LC material may be needed in order to achieve the calculated effective birefringence in an ultra-thin cell for a fast-response LCoS phase modulator. Thalhammer et al. utilized the overdrive approach to boost the LC response time to 1–2 ms for a low-resolution panel (256 × 256) with 500 Hz input frame rate and ~1.6 kHz data frame rate [39]. However, the input frame rate dropped to 30 Hz when using a High Definition Multimedia Iinterface (HDMI) controller in their high resolution (1920 × 1152) LCoS-SLM. Holoeye Photonics presented a HDMI 2.0 Field Programmable Gate Array (FPGA) driver to achieve 1920 × 1080 resolution (4000 ppi) for 240 Hz and 720 Hz data frame rates for high performance 3D sensing systems [40]. However, the driver scheme cannot support their LETO phase modulator (also 4000 ppi) due to the slow LC response time, which is not comparable with the sub-kHz input frame rates. Meadowlark Optics launched a new 1.8 ms response time phase-only 1920 × 1152 pixels (~2764 ppi) LCoS-SLM without overdrive. In 2018, the maximum frame rate reached has been 714 Hz by using a PCIe controller [41]. In the same year, Chen's team announced their PCU-3-01 LCoS-SLM, capable of reaching 1.6 ms at 45 °C. The panel can be driven at 240 Hz input frame rate with the OCM-ASIC SDK driver (Jasper Display Corp.) and at 720 Hz data frame rate by using an HDMI controller [17]. Higher resolution of 4K2K or 8K4K LCoS panels (greater than 5000 ppi) is limited by low input frame rate ($\leq$60 Hz) and low driving voltage (<3.3 V). Future electronic hardware updates are required to achieve full phase modulation. In summary, less than 2 ms LC response time is available for high-resolution LCoS-SLM panels, but the controller still needs to be updated to reach a frame rate over 500 Hz. The fast LC response of full phase modulation, high input frame rate, and driver interface are equally important to enable next-generation LCoS for holographic displays.

4. Phase Linearity

Phase linearity correction is an essential step for phase modulation panels. This is similar to the gamma correction for linear amplitude modulation in LC displays [42], but is a more complex process, covering the full 2 π radians phase. The linear phase response is a key criterion for satisfying computer-generated hologram (CGH) patterns calculated on the basis of the iterative Fourier transform algorithm (IFTA) or Fresnel (near-field)/Fraunhofer (far-field) diffraction equations, derived from scalar diffraction theory. The linear material for modulation is the main assumption of the theory. However, the intrinsic electro-optical response of liquid crystal is nonlinear. The measured intensity curve can be converted to phase based on following equation:

$$\frac{I}{I_0} = \cos^2 \chi - \sin 2\phi \sin 2(\phi - \chi) \sin^2 \left(\frac{\delta}{2} \right) \tag{1}$$

where I/I_0 is the normalized intensity, χ is the angle between the polarizer and the analyzer, ϕ is the angle between the LC director and the polarizer, and δ is the phase retardation. A nonlinear phase curve containing the sine or cosine function is tuned to a linear phase response using the look-up table (LUT) method [43]. The CGH simulation cannot correctly project its result without linear correction of LCoS-SLM. To solve this issue, "Linear Phase Calibration" is the first step for LCoS-SLM. The ease of calibration is closely related to the driving scheme in the control circuit. The digital driving scheme uses pulse width modulation (PWM) to generate the gray scale that can be divided into finer 0- and 1-bit planes. This makes programming of linear calibration much easier than the backplane using analog driving scheme. The average phase accuracy error (APAE%) and the root mean square (RMS) methods can be applied to evaluate the phase linearity of all gray levels:

$$\text{APAE\%} : \frac{\sum\limits_{GL=0}^{GL=255}\left(\frac{|\delta_{\text{m}}(LUT(GL))-\delta_i(GL)|}{\delta_{\max}-\delta_{\min}}\right)}{256} \tag{2}$$

$$\text{RMS} : \sqrt{\frac{\sum\limits_{GL=0}^{GL=255}\left(\delta_{\text{m}}(LUT(GL))-\delta_i(GL)\right)^2}{256}} \tag{3}$$

where GL is gray level, LUT is look-up table, and δ is the measured phase value. Let us take our PCM-2-01-633 LCoS-SLM panels with 1920 × 1080 resolution and 60–144 Hz picture frame rate as an example. A design with high-programmed pulse switching at ΔV = 5 V of the IC backplane was applied to reach a liquid crystal response within 4 ms. The phase linearity can be adjusted down to 1.08% (APAE%) and 0.024 π radians (RMS). In general, achieving ideal phase linearity in an analog driving LCoS-SLM is more difficult than in a digital driving scheme without additional optical compensation [44]. The Meadowlark Optics LCoS-SLM can achieve high-speed response (2–5 ms) while maintaining a certain degree of linearity in a low-resolution panel (256 × 256) using analog driving [45], but not in their higher resolution (HSP512 and HSP1920) LCoS panels. The LCoS panel with an analog drive from Hamamatsu Photonics can achieve ultra-high linearity (0.03 π radians (RMS)) at a lower resolution (792 × 600) and 60 Hz picture frame rate. However, to maintain ultra-high phase linearity, a thicker LC cell gap and lower operation voltage are required. The drawback of this setup is a slower response time (>30 ms).

5. Phase Precision

Besides the linear phase response in all gray scales, another issue is the phase precision over the entire active area. Phase precision has been discussed in many different forms, such as "spatially varying phase response" (SVPR) [46], "spatially resolved phase response" (SRPR) [47], "multipoint phase modulation" [48–51], "uniformity metrology" [52–55], "wavefront distortion" [56], "optical flatness", etc. The phase precision quality of an LCoS-SLM panel can be defined by using the above-mentioned evaluation methods. The current four major LCoS SLM manufacturers provide the following phase precision standards: (1) the latest 4K2K GAEA-2 from Holoeye Photonics—the optical flatness can be adjusted from $\lambda/6.6$ to $\lambda/10$ (RMS) after phase compensation; (2) the high phase precision LCoS panel from Hamamatsu Photonics—the phase precision has to be $\lambda/50$ mean standard deviations (mSTD) after phase compensation; (3) the latest HSP1920 from Meadowlark Optics—the phase precision can be adjusted from $\lambda/7$ to $\lambda/20$ (RMS); and (4) the latest SLM-200 from Santec Corp.—its wavefront distortion is greater than λ, but can be reduced to $\lambda/40$ after compensation. We adopt the mSTD from 256 (8 bits) gray scales to evaluate the phase precision before compensation, defined as:

$$mSTD(x,y) = \frac{\sum\limits_{GL=0}^{GL=255}\sqrt{\frac{\sum\limits_{mxy=1}^{mxy=\max}\left(\delta mxy(LUT_{global}(GL),x,y)-\overline{\delta mxy}(LUT_{global}(GL),x,y)\right)^2}{Total\ mxy}}}{256} \tag{4}$$

where δ is the phase measurement based on the LUT at different gray levels, and x and y refer to the specific area measured over the whole panel as shown in Figure 5. Fewer or more sections of mxy is dependent on the beam size or characterization method.

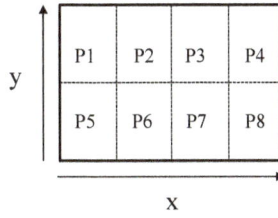

Figure 5. The scheme of eight different sessions (mxy) over the entire active area in the Equation (4).

The following formula represents various factors that affect phase modulation:

$$\delta_{\text{total}}(V,\ T,\ x,\ y) = \Delta\delta(T,\ V,\ x,\ y) + \delta_o(x,y) \tag{5}$$

Factors that can affect the output wavefront (δ_{total}) of an SLM can be attributed to modulated terms ($\Delta\delta$) and non-modulated terms (δ_o). The modulated terms ($\Delta\delta$) can be rewritten as follows to specify the effects on LC birefringence (Δn) and cell gap (d):

$$\delta_{\text{total}}(V,\ T,\ x,\ y) = \Delta n(T(x,\ y), V(x,y))d(x,y) + \delta_o(x,y) \tag{6}$$

In the non-modulated term, unevenness may arise due to the surface flatness of LCoS IC-backplane after the chemical mechanical polishing (CMP) process [57], the surface flatness of LC alignment layer, or other components like Indium Tin Oxide (ITO)cover glass. The main source of imprecision of the SLM phase is the modulation term. The three factors can be recognized across the entire LCoS panel (x, y) as following: (1) the cell gap uniformity (d); (2) the temperature effect (T) of the SLM panel, and (3) the voltage uniformity (V) from pixels. The uniformity of the LC layer is the most important parameter for the output wavefront. It requires a high assembly precision to control the cell gap uniformity error of ~1% over the entire active area in the liquid crystal assembly (LCA) process (assuming no internal damage to the IC backplane during LCA). This becomes more difficult if the cell gap is below 1.6 μm. The problem of rising temperature may be caused by either a high-power light source or the thermal effect from IC backplane. The output wavefront unevenness could result from non-uniform backplane temperature due to poor heat dissipation design in the optical system. This effect can be minimized if the operating temperature of the LCoS-SLM backplane can be evenly controlled at a constant temperature. Most of the problems that affect the phase precision of the panel are mainly due to the non-uniform LC cell gap and the uneven output voltage at each pixel. The non-uniform voltage occurs mainly on the analog drive's LCoS panel because the analog drive uses a dynamic random access memory (DRAM) capacitive method to supply different voltages for different gray levels. By contrast, different gray levels in the LCoS panel with digital drive are generated by the static random access memory (SRAM) to determine the voltage difference ($\Delta V = V_w - V_b$, in which V_w is V_{white}, and V_b is V_{black}) in conjunction with different time sequences (namely, pulse width) driving design. As a result, the digital driving scheme has the DC balance characteristic, which has fewer problems with voltage non-uniformity. In addition, the data addressing speed of the digital drive chip is very fast (ns ~ μs). This will not be the main cause of the unevenness and inaccuracy of the spatial phase, even if the digital driver adopts the time sequential driving scheme. The major problem associated with digital time sequential driving design is "phase flicker", which will be discussed in more detail in Section 6, which is devoted to phase stability.

The default setting may be off or not satisfied for the design applications after receiving the SLM panel from the venders. As a result, phase linearity calibration and phase precision compensation have been hot topics for SLM research. Various reported phase compensation methods can be roughly divided into two categories: one is by electronic compensation method, and the other is by optical compensation method. Most of the LCoS-SLM driver board settings are not open to users. As a result, the electronic compensation method may only be available to original manufacturers. Most SLM users applied optical algorithms to achieve high-quality wavefronts by correcting wavefront aberration using the Zernike polynomials algorithm. The algorithm can correct the most serious wavefront distortions, but is limited to a circular aperture due to the principle of Zernike polynomials. The compensation is inappropriate for square or rectangular apertures, such as the LCoS-SLM panel. For example, P512 series products from Boulder Nonlinear Systems (BNS), have wavefront distortion value of 1.2 λ before compensation. They can be optically compensated to λ/4 [58] but not up to the high phase precision standard (<λ/50). Their latest product, HSP1920, adopted electronic compensation for square sub-apertures in 2018. The wavefront distortion value can be greatly enhanced from λ/7 to λ/20 (RMS). Even though Zernike polynomials are not fully applicable to a square aperture to correct high-end aberrations, other research teams continue to improve or seek alternative optical algorithms for phase compensation. Xu et al. proposed to use the Zernike polynomials method plus least mean square fitting to adjust the LCoS uniformity of $1 \times 12{,}288$ pixels. Their result suggested that the RMS value of the inherent wave-front distortion can be suppressed approximately to λ/34 [59]. Engström et al. proposed using the principle of 7th-order 3D polynomials to compensate for the phase uniformity of the HSPDM512 product of BNS in the form of 64×64 pixels. It uses the peak-to-valley value as an evaluation method for wavefront distortion. It was shown that the maximum error amount can be corrected from 0.6–0.8 π to 0.3 π [46]. This method effectively corrects the error, but increases the overall computation time for the CGH phase compensation. The computation time can be accelerated up to 0.13 ms per CGH by using the Compute Unified Device Architecture (CUDA) encoding program. This result may not affect real-time operation on its panel, but will significantly prolong the computation time when using higher resolution (>4K2K) SLM with a high frame rate (>240 Hz). Either internal compensation using a chip circuit, or external compensation using an FPGA or Application Specific Integrated Circuit (ASIC) driver board are the better solutions to avoid extra load of computing the compensation algorithm in CGHs. In addition, it is possible to divide the active area into several unit blocks. The phase correction look-up table of each unit block (LUT(x, y)) can be performed individually to achieve high-precision modulation of the wavefront.

To demonstrate the phase precision measurements, we compare three LCoS-SLM at λ = 633 nm, PCM-2-01-633, PCU-3-01-633 (prepared in house), and LETO (Holoeye), under the same digital driving scheme at a controlled temperature T = 35 °C. The phase linearity errors (APAE%) under the global LUT of the PCM-2-01 and PCU-3-01 are 1.08% and 1.24%, respectively. The phase linearity error (APAE%) of the LETO panel can be tuned from 2.20% to 0.71% by using our driving calibration instead of the default setting for production. The spatially varying phase responses (SVPR) of the three different LCoS-SLM panels are measured after the phase linearity error of the global phase response is adjusted to around 1.0%. On each LCoS-SLM panel, a phase shift is estimated from eight different sections (P-1 to P-8 as shown in Figure 5) covering the entire panel's active area, as shown in Figure 6. The mSTD of phase precision error from PCM-2-01-633 and LETO are 0.035 π radians (λ/57) and 0.043 π radians (λ/47), respectively. For the PCU-3-01-633 panel, it is 0.067 π (λ/30). The cell gap uniformity errors of PCM-2-01-633 and PCU-3-01-633 panels are 0.68% and 0.99%, respectively, which are evaluated at 12 points using white light spectroscopy (Lambda 950 from Perkin Elmer). This rules out the panel thickness variation issue of these two panels. The problem of uniformity in data addressing is further evaluated by DAC (digital-to-analog converter) and BGP (background pixel) data driving modes. The measurement of DAC mode bypasses dynamic data addressing in pixel memory. It only fixed the IC backplane electrode voltage (V_0, V_1) and adjusted the voltage different of V_{ITO} ($V_{ITO-H} - V_{ITO-L}$). The DAC mode, in short, simply provides the V_{rms} of full panel as one large single pixel. The BGP mode is implanted in the linear global LUT data addressing in the SRAM. The phase precision errors from PCM-2-01-633 and PCU-3-01-633 panels are 0.035 and 0.039 π radians under DAC mode, but 0.035 and 0.067 π radians with BGP mode, respectively. It is clear that the non-uniform data addressing in pixel level is the cause of larger SVPR value in the PCU-3-01-633 LCoS-SLM panel.

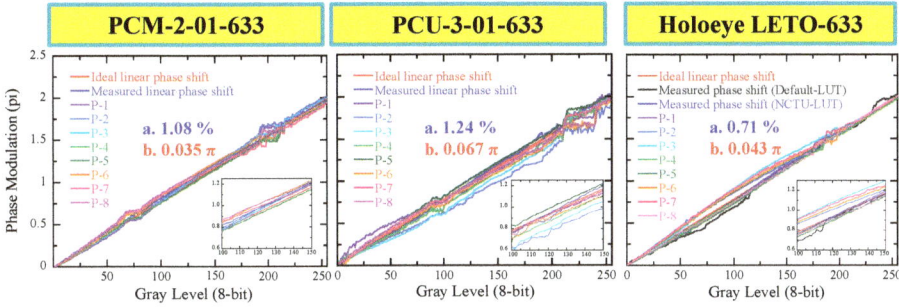

Figure 6. The phase precision measurements of 3 different LCoS-SLMs. The inserts are the enlarged figure from gray level 100 to 150. The values in a. is the phase linearity error (APAE%); and b. is the mSTD of phase precision error.

6. Phase Stability

Phase stability in terms of time-fluctuation is the major weakness in PWM devises [60–62]. The different working principles of digital and analog driving schemes are shown in Figure 7 [63].

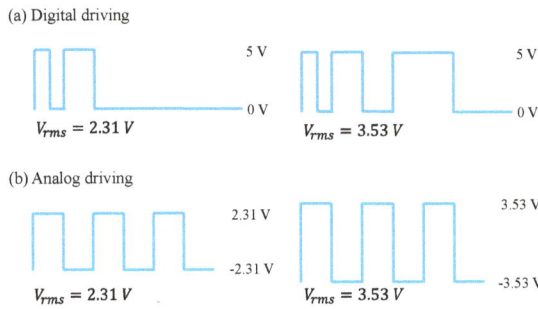

Figure 7. The schemes of digital and analog driving waveforms.

V_{rms} can be calculated between the voltages (v_1 and v_2) applied in a short period of time (t_1 and t_2), as shown in the following equation:

$$V_{rms} = \sqrt{\frac{t_1 v_1^2 + t_2 v_2^2}{t_1 + t_2}} \tag{7}$$

Briefly, the digital driving scheme maintains a constant voltage (ΔV) while changing the pulse width to achieve the various V_{rms} values. The flicker yields from charging and discharging phenomenon within the sub-millisecond because the instantaneous voltage recognized by liquid crystal is the time ratio between the pulses at constant voltage (ΔV) in the driving cycle. On the other hand, the analog driving scheme achieves its V_{rms} values by changing the applied voltages at constant frequency. There is no obvious flicker in the analog driving scheme because the instantaneous voltage resembling V_{rms} is perceived by the liquid crystals without a noticeable charging and discharging phenomenon. Figure 8 shows the phase flicker generated from digital drive at gray level 70 using a LETO panel.

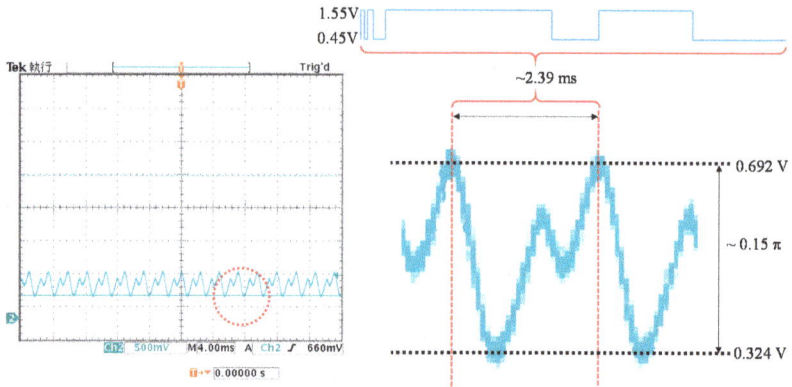

Figure 8. The digital drive waveform and its corresponding phase flicker measured in the high-resolution oscilloscope.

Two measurements with different photodetector settings are performed to show the phase shift and phase flicker in each gray scale. The digital color light sensor (TCS-3404 from TAOS) is adopted as a photodetector for the phase linearity and phase precision measurements. The data are taken every 100 ms for each gray level, which is presented as a red line for measured phase shift in Figures 9 and 10. Temporal fluctuation is measured by a polarization interferometer (PIF) with a polarizer and an analyzer aligned at 45 ° with respect to the rubbing direction. The photodetector is connected to a high-resolution oscilloscope (300 MHz 2.5 GS/s from Tektronix TDS3034B). The data are taken every few μs for each gray level. The phase flicker is determined by the maximum and minimum phase values of each gray level. The ΔV values of these three panels are 5.00V, 2.12V and 1.10V for PCM-2-01-633, PCU-3-01-633, and LETO, respectively. Their LC response times are 3.8, 2.8, and 20.7 ms, and their resulting mean standard deviations (mSTD) are 0.14 π, 0.12 π, and 0.04 π, respectively. The small ΔV setting in the digital driving scheme is able to minimize the flicker, as suggested in the data. To validate the assumption, the thicker cell gap of PCU-3-01-633 LCoS-SLM filled with the same LC was prepared to achieve 2π radius phase shift at lower ΔV = 1.04 V. The mSTD of the thicker LCoS-SLM (named PCU-3-01-633-2) is only 0.04 π. The response time, regrettably, slows considerably from 2.8 to 9.6 ms. To summarize the phase stability in the digital drive LCoS-SLM, it is able to minimize the flicker when smaller ΔV is applied for 2 π radius phase shift modulation. The slower LC response, however, is inevitable with this approach.

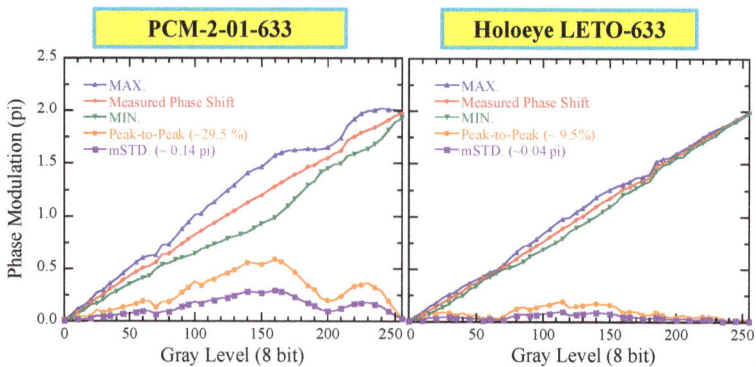

Figure 9. The phase stability measurements of PCM-2-01-633, and LETO LCoS-SLMs at 633 nm.

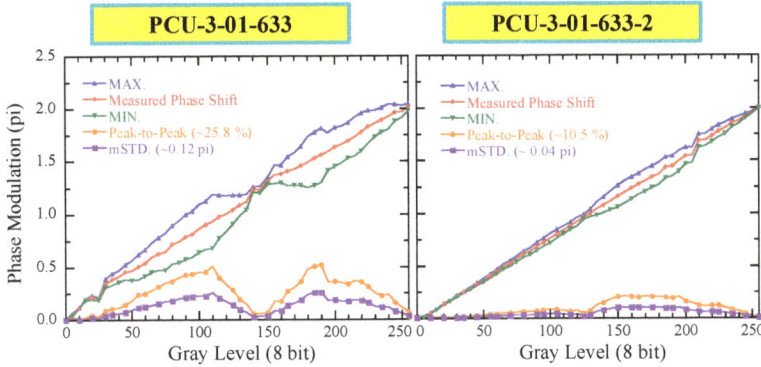

Figure 10. The phase stability measurements of PCU-3-01-633 with two different cell gap LCoS-SLMs at 633 nm.

7. Phase Accuracy

Phase accuracy is determined by the projected image. The pixel-level spatially anamorphic phenomenon provides a way to appraise the phase accuracy from a linear phase calibrated panel [64–67]. The phase error is quantified by employing various spatial frequencies and grating directions deviating from an ideal crenel-like binary grating. The deviation from the ideal binary grating is related to the non-desired LC fringing field and reverse-tilt effects [67]. To measure phase modulation, a 0th gray level (GL-0) is defined as 100% of the intensity of the light reflected from the panel. The ±1st-order light intensity is determined by the spatial frequency and orientation of the binary grating. The binary-grating differences are created by changing the corresponding gray-level pairs, in which the gray level of the reference pixel is fixed as GL-0 and changed 5–255 gray levels (GL-5-255) in the counterpart. The PCM-2-01-633, PCU-3-01-633, and LETO LCoS-SLM panels, possessing the same backplane design, are compared regarding their pixel-level spatially anamorphic phenomenon at λ = 633 nm. In Figure 11, H and V represented horizontal and vertical grating, and P is the binary grating pairs at different periods: 1 + 1, 2 + 2, and 6 + 6 pixels/grating, which correspond to spatial frequencies of 76.8, 38.4, and 12.8 lines/mm, respectively. The average error of the pixel-level phase accuracy of the horizontal and vertical grating is 9% for PCM-2-01-633, 7.7% for PCU-3-01-633, and 12% for LETO. This result suggests that PCU-3-01-633 with a small cell gap can effectively suppress the pixel-level phase accuracy error. However, the 1 + 1 in vertical grating still has 10% error. Small average phase-accuracy errors in both horizontal and vertical grating directions are necessary at the same time to minimize the error for the application of 2D CGH patterns.

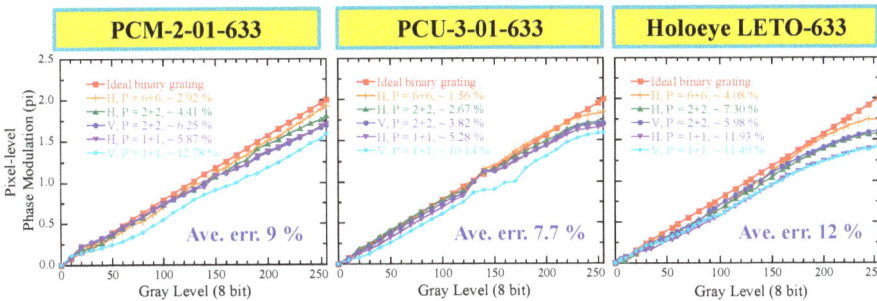

Figure 11. The phase accuracy measurements of three different LCoS-SLMs at 633 nm.

8. Conclusions

LCoS technology has been developed for nearly four decades. However, due to its higher production cost and lower performance than DMD, the latter continues to dominating the present-day projection display market. Recently, high-resolution and fine pixel size LCoS backplanes have been produced by several vendors, and appear to offer more promising spatial light modulation than full high definition displays. Techniques for assessing various characteristics of high-resolution LCoS SLMs have been discussed in this review. The physics and modeling of LCs in parallel aligned LCoS were not discussed in detail due to the limited length and scope of this paper. The physics and electro-optical effects of tilted LCs, and addressed digital driving in the parallel aligned LCoS, have been thoroughly discussed by Francés et al. and Martínez, et al. [68–70]. During the course of developing high quality SLM panels, the phase precision, phase stability, and output phase accuracy need to be meticulously adjusted on a linearly calibrated phase panel. The novel high-quality and well-calibrated panels can generate phase-only holography and project faithful CGH design, which can be useful for 2D/3D holographic video projectors and optical telecommunications. The high phase precision, fast response time, good phase stability, and high phase accuracy of LCoS panels opens more possibilities for various applications, as discussed above. Though the benefits of LCoS panels are many, endeavors should be made to offer them at a competitive price to the public in the near future.

Author Contributions: Conceptualization: H.-M.P.C. and J.-P.Y.; writing—original draft preparation, J.-P.Y. and H.-M.P.C.; writing—review and editing, H.-M.P.C. and S.-T.W.; experimental work, J.-P.Y., H.-T.Y., Z.-N.H. and Y.H.; supervision, H.-M.P.C. and S.-T.W.

Funding: The financial support is provided by the Ministry of Science and Technology of the Republic of China under Grant Nos. 105-2622-E-009-012-CC2, 105-2221-E-009-090, and MOST 106-2221-E-009-111; and is partially supported by the Research Team of Photonic Technologies and Intelligent Systems at NCTU within the framework of the Higher Education Sprout Project by the Ministry of Education (MOE) in Taiwan.

Acknowledgments: The authors are grateful for the LCoS photos listed in Figure 1 provided by Jasper Display.

References

1. Margerum, J.D.; Nimoy, J.; Wong, S.-Y. Reversible ultraviolet imaging with liquid crystals. *Appl. Phys. Lett.* **1970**, *17*, 51–53. [CrossRef]
2. Beard, T.D.; Bleha, W.P.; Wong, S.-Y. Ac liquid-crystal light valve. *Appl. Phys. Lett.* **1973**, *22*, 90–92. [CrossRef]
3. Ernstoff, M.N.; Leupp, A.M.; Little, M.J.; Peterson, H.T. Liquid crystal pictorial display. *Int. Electron. Devices Meet.* **1973**, 548–551. [CrossRef]
4. Efron, U.; Braatz, P.O.; Little, M.J.; Schwartz, R.N.; Grinberg, J. Silicon liquid crystal light valves: Status and issues. *Opt. Eng.* **1983**, *22*, 682–686. [CrossRef]
5. Efron, U.; Wu, S.T.; Bates, T.D. Nematic liquid crystals for spatial light modulators: Recent studies. *J. Opt. Soc. Am. B.* **1986**, *3*, 247–252. [CrossRef]
6. Johnson, K.M.; McKnight, D.J.; Underwood, I. Smart Spatial Light Modulators Using Liquid Crystals on Silicon. *IEEE J. Quantum Electron.* **1993**, *29*, 699–714. [CrossRef]
7. Vettese, D. Liquid crystal on silicon. *Nat. Photonics* **2010**, *4*, 752–754. [CrossRef]
8. Maimone, A.; Georgiou, A.; Kollin, J.S. Holographic Near-Eye Displays for Virtual and Augmented Reality. *ACM Trans. Graph.* **2017**, *36*, 85. [CrossRef]
9. Matsuda, N.; Fix, A.; Lanman, D. Focal Surface Displays. *ACM Trans. Graph.* **2017**, *36*, 86. [CrossRef]
10. Lazarev, G.; Gädekea, F.; Luberek, J. Ultrahigh-resolution phase-only LCOS spatial light modulator. In Proceedings of the Emerging Liquid Crystal Technologies XII, San Francisco, CA, USA, 28 January–2 February 2017.
11. Inoue, T.; Tanaka, H.; Fukuchi, N.; Takumi, M.; Matsumotoa, N.; Hara, T.; Yoshida, N.; Igasaki, Y.; Kobayashi, Y. LCOS spatial light modulator controlled by 12-bit signals for optical phase-only modulation. In Proceedings of the Emerging Liquid Crystal Technologies II, San Jose, CA, USA, 9 February 2007.
12. Linnenberger, A. Advanced SLMs for Microscopy. *Proc. SPIE* **2018**, *10502*, 1050204.
13. Santec. Available online: http://www.santec.com/en/products/components/slm (accessed on 29 October 2018).

14. Wang, M.; Zong, L.; Mao, L.; Marquez, A.; Ye, Y.; Zhao, H.; Vaquero, F.J. LCoS SLM Study and Its Application in Wavelength Selective Switch. *Photonics* **2017**, *4*, 22. [CrossRef]

15. Vizsnyiczai, G.; Kelemen, L.; Ormos, P. Holographic multi-focus 3D two-photon polymerization with real-time calculated holograms. *Opt. Express* **2014**, *22*, 24217–24223. [CrossRef] [PubMed]

16. Beck, R.J.; Parry, J.P.; Shephard, J.D.; Hand, D.P. Compensation for time fluctuations of phase modulation in a liquid-crystal-on-silicon display by process synchronization in laser materials processing. *Appl. Opt.* **2011**, *50*, 2899–2905. [CrossRef] [PubMed]

17. Toyoda, H.; Inoue, T.; Mukozaka, N.; Hara, T.; Wu, M.H. Advances in Application of Liquid Crystal on Silicon Spatial Light Modulator (LCOS-SLM). *SID Int. Symp. Dig. Tech.* **2014**, *45*, 559–562. [CrossRef]

18. Vieri, C.; Lee, G.; Balram, N.; Jung, S.H.; Yang, J.Y.; Yoon, S.Y.; Kang, I.B. An 18 megapixel 4.3" 1443 ppi 120 Hz OLED display for wide field of view high acuity head mounted displays. *J. Soc. Inf. Disp.* **2018**, *26*, 314–324. [CrossRef]

19. Lee, H.S.; Jang, S.; Jeon, H.; Choi, B.S.; Cho, S.H.; Kim, W.T.; Song, K.; Chu, H.Y.; Kim, S.; Jo, S.C.; et al. Large-area Ultra-high Density 5.36" 10Kx6K 2250 ppi Display. *Proc. SID Symp. Dig. Tech.* **2018**, *49*, 607–609. [CrossRef]

20. Wakunami, K.; Hsieh, P.Y.; Oi, R.; Senoh, T.; Sasaki, H.; Ichihashi, Y.; Okui, M.; Huang, Y.P.; Yamamoto, K. Projection-type see-through holographic three-dimensional display. *Nat. Commun.* **2016**, *7*, 12954. [CrossRef] [PubMed]

21. Kim, Y.H.; Hwang, C.Y.; Choi, J.H.; Pi, J.E.; Yang, J.H.; Cho, S.M.; Cheon, S.H.; Kim, G.H.; Choi, K.; Kim, H.O.; et al. Development of high-resolution active matrix spatial light modulator. *Opt. Eng.* **2018**, *57*, 061606. [CrossRef]

22. Holoeye. Available online: https://holoeye.com/spatial-light-modulators/gaea-4k-phase-only-spatial-light-modulator/ (accessed on 29 October 2018).

23. Yang, J.P.; Chen, H.M.P.; Huang, Y.; Wu, S.T.; Hsu, C.; Ting, L.; Hsu, R. Sub-KHz 4000-PPI LCoS Phase Modulator for Holographic Displays. *Proc. SID Symp. Dig. Tech.* **2018**, *49*, 772–775. [CrossRef]

24. Abeeluck, A.K.; Iverson, A.; Goetz, H.; Passon, E. High-Performance Displays for Wearable and HUD Applications. *Proc. SID Symp. Dig. Tech.* **2018**, *49*, 768–771. [CrossRef]

25. Zeng, Z.; Zheng, H.; Yu, Y.; Asund, A.K.; Valyukh, S. Full-color holographic display with increased-viewing-angle. *Appl. Opt.* **2017**, *56*, F112–F120. [CrossRef] [PubMed]

26. Yang, J.P.; Chen, H.M.P. A 3-msec Response-Time Full-Phase-Modulation 1080p LCoS-SLM for Dynamic 3D Holographic Displays. *Proc. SID Symp. Dig. Tech.* **2017**, *48*, 1073–1076. [CrossRef]

27. Inoue, T.; Takaki, Y. Table screen 360-degree holographic display using circular viewing-zone scanning. *Opt. Express* **2015**, *23*, 6533–6542. [CrossRef] [PubMed]

28. Ko, S.B.; Park, J.H. Speckle reduction using angular spectrum interleaving for triangular mesh based computer generated hologram. *Opt. Express* **2017**, *25*, 29788–29797. [CrossRef] [PubMed]

29. Fourth Dimension Displays. Available online: https://www.forthdd.com/products/spatial-light-modulators/ (accessed on 29 October 2018).

30. Lin, C.W.; Chen, H.M.P. Defect-free half-V-mode ferroelectric liquid-crystal device. *J. Soc. Inf. Disp.* **2010**, *18*, 976–980. [CrossRef]

31. Turtaev, S.; Leite, I.T.; Mitchell, K.J.; Padgett, M.J.; Phillips, D.B.; Čižmár, T. Comparison of nematic liquid-crystal and DMD based spatial light modulation in complex photonics. *Opt. Express* **2017**, *25*, 29874–29884. [CrossRef] [PubMed]

32. Xianyu, H.; Wu, S.T.; Lin, C.L. Dual frequency liquid crystals: A review. *Liq. Cryst.* **2009**, *36*, 717–726. [CrossRef]

33. Peng, F.; Xu, D.; Chen, H.; Wu, S.T. Low voltage polymer network liquid crystal for infrared spatial light modulators. *Opt. Express* **2015**, *23*, 2361–2368. [CrossRef] [PubMed]

34. Sun, J.; Chen, Y.; Wu, S.T. Submillisecond-response and scattering-free infrared liquid crystal phase modulators. *Opt. Express* **2012**, *20*, 20124–20129. [CrossRef] [PubMed]

35. Peng, F.; Lee, Y.H.; Luo, Z.; Wu, S.T. Low voltage blue phase liquid crystal for spatial light modulators. *Opt. Lett.* **2015**, *40*, 5097–5100. [CrossRef] [PubMed]

36. Huang, Y.; He, Z.; Wu, S.T. Fast-response liquid crystal phase modulators for augmented reality displays. *Opt. Express* **2017**, *25*, 32757–32766. [CrossRef]

37. Chen, R.; Huang, Y.; Li, J.; Hu, M.; Li, J.; Wu, S.T.; An, Z. High-frame-rate liquid crystal phase modulator for augmented reality displays. *Liq. Cryst.* **2018**. [CrossRef]
38. Wu, S.T.; Efron, U. Optical properties of thin nematic liquid crystal cells. *Appl. Phys. Lett.* **1986**, *48*, 624–626. [CrossRef]
39. Thalhammer, G.; Bowman, R.W.; Love, G.D.; Padgett, M.J.; Ritsch-Marte, M. Speeding up liquid crystal SLMs using overdrive with phase change reduction. *Opt. Express* **2013**, *21*, 1779–1797. [CrossRef] [PubMed]
40. Lazarev, G.; Bonifer, S.; Engel, P.; Höhne, D.; Notni, G. High-resolution LCOS microdisplay with sub-kHz frame rate for high performance, high precision 3D sensor. *Proc. SPIE Dig. Opt. Technol.* **2017**, *10335*. [CrossRef]
41. Meadowlark Optics. Available online: https://www.meadowlark.com/1920-1152-spatial-light-modulator-p-119?mid=18#.W34yX84zYdU (accessed on 29 October 2018).
42. Xiao, K.D.; Fu, C.Y.; Dimosthenis, K.; Wuerger, S. Visual gamma correction for LCD displays. *Displays* **2011**, *32*, 17–23. [CrossRef]
43. Yang, L.; Xia, J.; Chang, C.; Zhang, X.; Yang, Z.; Chen, J. Nonlinear dynamic phase response calibration by digital holographic microscopy. *Appl. Opt.* **2015**, *54*, 7799–7806. [CrossRef] [PubMed]
44. Strauß, J.; Häfner, T.; Dobler, M.; Heberle, J.; Schmidt, M. Evaluation and calibration of LCoS SLM for direct laser structuring with tailored intensity distributions. *Phys. Procedia* **2016**, *83*, 1160–1169. [CrossRef]
45. Zhang, H.; Zhou, H.; Li, J.; Qiao, Y.J.; Si, J.; Gao, W. Compensation of phase nonlinearity of liquid crystal spatial light modulator for high-resolution wavefront correction. *J. Eur. Opt. Soc.-Rapid* **2015**, *10*, 15036. [CrossRef]
46. Engström, D.; Persson, M.; Bengtsson, J.; Goksör, M. Calibration of spatial light modulators suffering from spatially varying phase response. *Opt. Express* **2013**, *21*, 16086–16103. [CrossRef] [PubMed]
47. Reichelt, S. Spatially resolved phase-response calibration of liquid-crystal-based spatial light modulators. *Appl. Opt.* **2013**, *52*, 2610–2618. [CrossRef] [PubMed]
48. Otón, J.; Ambs, P.; Millán, M.S.; Cabré1, E.P. Multipoint phase calibration for improved compensation of inherent wavefront distortion in parallel aligned liquid crystal on silicon displays. *Appl. Opt.* **2007**, *46*, 5667–5679. [CrossRef] [PubMed]
49. Otón, J.; Ambs, P.; Millán, M.S.; Cabré1, E.P. Dynamic calibration for improving the speed of a parallel-aligned liquid-crystal-on-silicon display. *Appl. Opt.* **2009**, *48*, 4616–4624. [CrossRef] [PubMed]
50. Lu, Q.; Sheng, L.; Zeng, F.; Gao, S.; Qiao, Y. Improved method to fully compensate the spatial phase nonuniformity of LCoS devices with a Fizeau interferometer. *Appl. Opt.* **2016**, *55*, 7796–7802. [CrossRef] [PubMed]
51. Xia, J.; Chang, C.; Chen, Z.; Zhu, Z.; Zeng, T.; Liang, P.Y.; Ding, J. Pixel-addressable phase calibration of spatial light modulators: A common-path phase-shifting interferometric microscopy approach. *J. Opt.* **2017**, *19*, 125701. [CrossRef]
52. Gelder, R.V.; Melnik, G. Uniformity metrology in ultra-thin LCoS LCDs. *J. Soc. Inf. Disp.* **2006**, *14*, 233–239. [CrossRef]
53. Zhang, Z.; Chapman, A.M.J.; Collings, N.; Pivnenko, M.; Moore, J.; Crossland, B.; Chu, D.P.; Milne, B. High Quality Assembly of Phase-Only Liquid Crystal on Silicon (LCOS) Devices. *J. Disp. Technol.* **2011**, *7*, 120–126. [CrossRef]
54. Zhang, Z.; Yang, H.; Robertson, B.; Redmond, M.; Pivnenko, M.; Collings, N.; Crossland, W.A.; Chu, D.P. Diffraction based phase compensation method for phase-only liquid crystal on silicon devices in operation. *Appl. Opt.* **2012**, *51*, 3837–3846. [CrossRef] [PubMed]
55. Zhang, Z.; Pivnenko, M.; Salazar, I.M.; You, Z.; Chu, D.P. Advanced die-level assembly techniques and quality analysis for phase-only liquid crystal on silicon devices. *Proc IMechE Part B J Eng. Manuf.* **2016**, *230*, 1659–1664. [CrossRef]
56. Čižmár, T.; Mazilu, M.; Dholakia, K. In situ wavefront correction and its application to micromanipulation. *Nat. Photonics* **2010**, *4*, 388–394. [CrossRef]
57. Serati, S.; Xia, X.; Mughal, O.; Linnenberger, A. High-resolution phase-only spatial light modulators with submillisecond response. In Proceedings of the Optical Pattern Recognition XIV, Orlando, FL, USA, 6 August 2003.

58. Harriman, J.; Linnenberger, A.; Serati, S. Improving spatial light modulator performance through phase compensation. In Proceedings of the Advanced Wavefront Control: Methods, Devices, and Applications II, Denver, CO, USA, 12 October 2004.

59. Xu, J.; Qin, S.; Liu, C.; Fu, S.; Liu, D. Precise calibration of spatial phase response nonuniformity arising in liquid crystal on silicon. *Opt. Lett.* **2018**, *43*, 2993–2996. [CrossRef] [PubMed]

60. Collings, N.; Christmas, J.L. Real-Time Phase-Only Spatial Light Modulators for 2D Holographic Display. *J. Disp. Technol.* **2015**, *11*, 278–284. [CrossRef]

61. Lizana, A.; Márquez, A.; Lobato, L.; Rodange, Y.; Moreno, I.; Iemmi, C.; Campos, J. The minimum Euclidean distance principle applied to improve the modulation diffraction efficiency in digitally controlled spatial light modulators. *Opt. Express* **2010**, *18*, 10581–10593. [CrossRef] [PubMed]

62. Márquez, J.G.; López, V.; Vega, A.G.; Noél, E. Flicker minimization in an LCoS spatial light modulator. *Opt. Express* **2012**, *20*, 8431–8441. [CrossRef] [PubMed]

63. Moore, J.R.; Collings, N.; Crossland, W.A.; Davey, A.B.; Evans, M.; Jeziorska, A.M.; Komarcevic, M.; Parker, R.J.; Wilkinson, T.D.; Xu, H. The Silicon Backplane Design for an LCOS Polarization-Insensitive Phase Hologram SLM. *IEEE Photonics Technol. Lett.* **2008**, *20*, 60–62. [CrossRef]

64. Bouvier, M.; Scharf, T. Analysis of nematic-liquid-crystal binary gratings with high spatial frequency. *Opt. Eng.* **2000**, *39*, 2129–2137. [CrossRef]

65. Márquez, A.; Iemmi, C.; Moreno, I.; Campos, J.; Yzuel, M.J. Anamorphic and spatial frequency dependent phase modulation on liquid crystal displays. Optimization of the modulation diffraction efficiency. *Opt. Express* **2005**, *13*, 2111–2119. [CrossRef] [PubMed]

66. Cuypers, D.; Smet, H.D.; Calster, A.V. Electronic Compensation for Fringe-Field Effects in VAN LCOS Microdisplays. *SID Symp. Dig. Tech.* **2008**, *Papers 39*, 228–231. [CrossRef]

67. Lobato, L.; Lizana, A.; M'arquez, A.; Moreno, I.; Iemmi, C.; Campos, J.; Yzuel, M.J. Characterization of the anamorphic and spatial frequency dependent phenomenon in liquid crystal on silicon displays. *J. Eur. Opt. Soc.* **2011**, *6*, 11012S. [CrossRef]

68. Francés, J.; Márquez, A.; Martínez-Guardiola, F.J.; Bleda, S.; Gallego, S.; Neipp, C.; Pascual, I.; Beléndez, A. Simplified physical modeling of parallel-aligned liquid crystal devices at highly non-linear tilt angle profiles. *Opt. Express* **2018**, *26*, 12723–12741. [CrossRef] [PubMed]

69. Martínez, F.J.; Márquez, A.; Gallego, S.; Francés, J.; Pascual, I.; Beléndez, A. Effective angular and wavelength modeling of parallel aligned liquid crystal devices. *Opt. Laser Eng.* **2015**, *74*, 114–121. [CrossRef]

70. Martínez, F.J.; Márquez, A.; Gallego, S.; Francés, J.; Ortuño, M.; Francés, J.; Beléndez, A.; Pascual, I. Electrical dependencies of optical modulation capabilities in digitally addressed parallel aligned liquid crystal on silicon devices. *Opt. Eng.* **2014**, *53*, 067104.

applied
sciences

MDPI

Review

Progress in Phase Calibration for Liquid Crystal Spatial Light Modulators

Rujia Li and Liangcai Cao *

State Key Laboratory of Precision Measurement Technology and Instrument, Department of Precision Instruments, Tsinghua University, Beijing 100084, China; lrj18@mails.tsinghua.edu.cn
* Correspondence: clc@tsinghua.edu.cn

Received: 4 April 2019; Accepted: 10 May 2019; Published: 16 May 2019

Abstract: Phase-only Spatial Light Modulator (SLM) is one of the most widely used devices for phase modulation. It has been successfully applied in the field with requirements of precision phase modulation such as holographic display, optical tweezers, lithography, etc. However, due to the limitations in the manufacturing process, the grayscale-phase response could be different for every single SLM device, even varying on sections of an SLM panel. A diverse array of calibration methods have been proposed and could be sorted into two categories: the interferometric phase calibration methods and the diffractive phase calibration methods. The principles of phase-only SLM are introduced. The main phase calibration methods are discussed and reviewed. The advantages of these methods are analyzed and compared. The potential methods for different applications are suggested.

Keywords: phase measurement; spatial light modulator; calibration; interference; diffraction

1. Introduction

Liquid Crystal Spatial Light Modulator (LC-SLM) and Digital Micro-Mirror device (DMD) are two of the most successful SLMs. For phase manipulation, the LC-SLM has been widely used in many fields: holographic display [1–3], precision metering [4,5], optical tweezers [6,7], lithography [8], beam shaping [9,10], etc. In the listed fields, the accurate phase manipulation is essential, especially for deploying SLM as a programmable diffractive device. The shifted phase could be modulated by the addressed grayscale or pixel voltage for an electric-addressed SLM. Ideally, the transfer from the uploaded grayscale to the shifted phase of the light wave is the same as the look up table (LUT) shown in the startup manual of the SLM. However, in practice, due to the nonlinear optical response of liquid crystal as well as manufacturing defects, the grayscale-phase response could be different for every single SLM device, even varying on sections of an SLM panel.

The performance reduction of SLM is due to two kinds of static errors: the global grayscale-phase mismatch and spatial nonuniformity. The global grayscale-phase mismatch comes from the application of the unsuitable pre-set LUT of the SLM. The spatial nonuniformity of SLM's active area (on the panel to modulate incident beam) comes from the nonuniform electric drive [11], the curvature of the backplane of the SLM [12], or the aberration induced by the thickness variations of the LC layer. Besides, the accuracy of the applied electric field transferred from the uploaded grayscale for the pixels could lead to errors. For different illuminating wavelengths, the response will be different. Along with the static errors, there could be fluctuations of the phase modulation caused by the dynamic errors, including the local heating caused by an incident laser beam and the changes of environmental condition, such as temperature or humidity. Besides, performance reduction could happen as the SLM is aging. Therefore, the phase modulation of the SLM will not be the same as the way it is designed, and phase modulation calibration is needed for improving the performance of the SLM involved system, especially the static calibration [13]. For instance, with spatially calibrated SLM, the control of the trapping intensities could increase in the holographic optical trapping system [11].

The key point of the SLM calibration process is to acquire the phase corresponding to the grayscale. The main task is to measure the modulated phase of the SLM efficiently and accurately. All of the phase calibration methods are aiming to transfer the indirectly detectable phase distributions to detectable intensities such as interference fringes that could be used for analysis. A diverse array of calibration methods have been proposed and could be sorted into two categories: the interferometric phase calibration method and the diffractive phase calibration method. Both these methods provide solutions for transforming the phase to intensities, which are also widely applied in the measurements of all kinds of phase objects, such as the quantitative phase imaging for biological sample characterization.

In this work, the LC-SLM and LCoS are investigated for instance to review the progress in the calibration methods. The interferometric methods are firstly discussed, which contain the self-referenced and out-referenced calibration methods. The diffractive methods are then described, which contain the diffraction pattern analysis method, optical elements generated method and polarization analyzed method. The advantages of these methods are analyzed and compared. The potential methods for different applications are suggested. The suggestions could be a reference for the calibrations of any SLM for its applications.

2. Basic Principle of Phase-Only SLM

The liquid crystal refers to a state of matter between the liquid and solid crystal. There are many phases of liquid crystal. The nematic and smectic C * ferroelectric phases of LCs are the most popular for SLMs. Currently, the ferroelectric LC-based SLM could reach binary phase modulation with a very rapid response, and the SLM based on it could lead to obvious zeroth order and conjugated images at the same time. For the multi-grayscale phase modulation, the nematic phases of LC are suitable with reasonable viscosity and response.

The SLM are like sandwiches with different layers. Taking the reflective LCoS for instance, the basic structure of reflective SLM can be shown in Figure 1a. As it is called Liquid Crystal on Silicon, the LC layer is arranged on the silicon substrate. Based on the semiconductor technology, the active matrix circuit is formed on the silicon substrate to control each pixel electrode by the applied electrical potential individually. The pixel electrode could be made by reflective aluminum with pixel circuity, which contains gate lines and buried transistors for a maximum fill factor [14,15]. Two alignment layers with rubbing direction could be used to initialize the orientation of the modulated LC molecules. The transparent electrode is deployed to work with the pixel electrode for generating the electric field across the LC cell. The birefringence could be modulated by the electric field, which is an electrically controlled birefringence (ECB) mode.

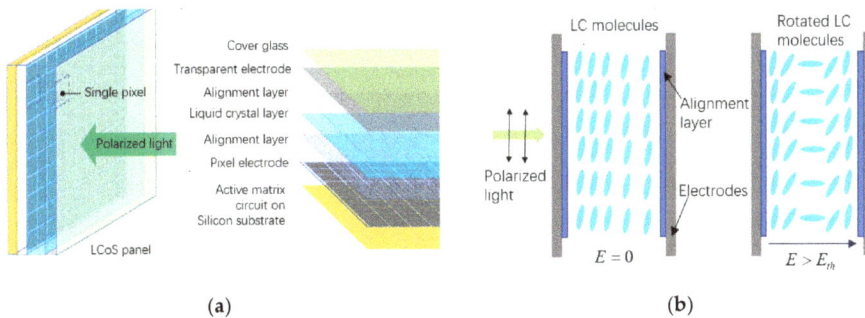

Figure 1. (a) Basic structure of LCoS. (b) Schematic of zero-twisted ECB mode-based LC cell in ON and OFF states. LC: liquid crystal; E_{th}: threshold of the electric field amplitude.

For optical applications, the director of the LC molecules plays an essential role in the birefringence. Before phase modulation, the incident light should be polarized for matching the director of LC. The zero-twisted nematic ECB mode is one of the most widely used phase modulation modes for LCoS. As shown in Figure 1b, the electric field could be generated by the electrodes. The alignment layers provide a pre-orientation of the small angle of 2° to prevent reverse tilt switching [16]. The LC molecules are homogeneously quasi-parallel (with 2°) aligned to the electrode panel when no electric field is applied. When the electric field is applied and the amplitude is larger than the threshold of the electric field amplitude [17], the LC molecules will tilt. The effective refractive index of the cell varies with the molecules polar angle $\theta(E)$ related to the applied electric field. Thus, the phase retardation could be modulated by changing the applied electric field. Assuming complete polarization and no absorption, the Jones matrix of the LC cell could be:

$$J_{ECB} = \begin{pmatrix} 1 & 0 \\ 0 & \exp[-i\Gamma(E)] \end{pmatrix} \qquad (1)$$

The phase retardation $\Gamma(E)$ could be:

$$\Gamma(E) = \frac{2\pi}{\lambda} d\big(n_{eff}(E) - n_o\big) \qquad (2)$$

where λ is the wavelength, d is the thickness of the LC layer, $n_{eff}(E)$ is the effective refractive index varying with the applied electric field, and n_o is the ordinary refractive index of the LC [18]. The $n_{eff}(E)$ ranges from n_o to n_e (extraordinary refractive index), which depends on the applied electric field. The birefringence $\Delta n = n_e - n_o$ of positive LC could be from 0.05 to 0.45 [19,20].

The pattern with the grayscale values uploaded on SLM could be transferred to the applied electric field on the pixel electrodes to control the phase modulation of each pixel. Ideally, the grayscale-phase relation set to the commercial phase-only SLM should be linear, and the gamma curve of the ideal SLM is shown as the red line in Figure 2a. In fact, with the aforementioned imperfectness, the shifted phase modulated by SLM without calibration is represented by the red lines in Figure 2b. For linear modulations of the phase described by green lines in Figure 2b, the desired gamma curve can be shown as the green curve in Figure 2a. In this case, a necessary calibration is desired.

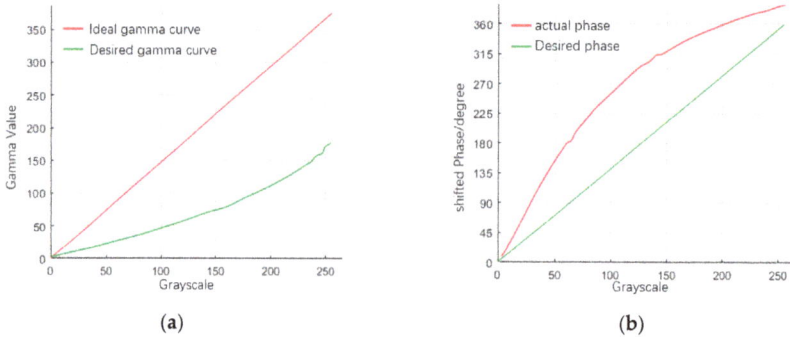

Figure 2. (**a**) The pre-set ideal gamma curve (red) and the desired gamma curve (green) for linear phase modulation. (**b**) The shifted phase modulated by SLM without calibration (red) and the desired linear shifted phase (green) [21].

For the phase modulation of SLM, there is a response time for the molecules in LC cells to be tilted to the orientation needed for modulation. The response could be analyzed from anchoring effects and cell gap effects. The anchoring effects refer to the anisotropic part of the interfacial free energy for the alignment layer [22]. The cell gap refers to the thickness of the LC layer. The response time is proportional to d^x, where x is related to the anchoring energy and d is the cell gap [23]. Assuming strong anchoring effects, the reflective LCoS could respond four times faster than the transmissive LC-SLM for the same phase retardation. For the LCoS, it takes milliseconds to refresh a grayscale image.

The SLMs are pixelated digital devices consisting of millions of pixels (such as the resolution of 1920×1080). For each pixel, the pixel size and fill factor are two key parameters. For instance, the pixel size (pixel pitch) of commercial SLM could be 3.74 μm × 3.74 μm so that the highest spatial frequency of the image displayed on the SLM is limited. A smaller pixel pitch leads to better resolution and larger diffraction angles. The fill factor influences the intensity of the zeroth order in the modulated diffracted beam [24]. The zeroth order could disturb the image as well as the calibration procedure so that the SLM with a high fill factor is preferred. Even when the fill factor is 100%, the zeroth order cannot be eliminated because of the crosstalk [25].

For high-precision applications of the SLM, the properly polarized incident beam with an illumination angle could lead to a reduction or increase of the phase modulation. Besides, illuminating anisotropic LC with a large angle could lead to multi-reflection on the surface of the layer of LC-SLM or LCoS, in which the condition of the unwanted additional phase related to the addressed grayscale could appear [26]. Hence, normal illumination with proper polarization is the most favorable for phase-only LC-SLM while uploading a grayscale image of complexity, such as the application of holographic display. Alternatively, the incident angle for application could be held as the same as the illumination angle for the calibration procedure if the phase distortion is tolerable. Along with the illumination angle, the power of the incident laser could cause local heating. The phase modulation characteristic could vary with incident power so that the calibration could be carried out under the same incident laser power as the application to minimize the local heating effects [11]. Besides, the threshold of the damage power of the SLM nowadays could be 2 W/cm^2.

3. Interferometric Phase Calibration Method

The SLM phase calibration could be treated as a modulated phase detection process with a foregone changing grayscale pattern uploaded on SLM. Due to the extremely low detection rate of the electrical photo-electric detector compared to the frequency of the light, directly detecting the phase of the light is an impossible mission using an available detector which could record the intensity only. The interferometric technique could be utilized to record and reconstruct both the intensity and phase by mathematical methods. For instance, the interference fringes of two monochromatic plane waves with the same direction of polarization could be described as:

$$I(x,y) = I_1(x,y) + I_2(x,y) + 2\sqrt{I_1(x,y) \cdot I_2(x,y)} \cdot \cos\delta \tag{3}$$

where $I_1(x,y)$ and $I_2(x,y)$ are the intensity distributions on the detection plane of two beams, δ is the phase difference between the two beams, and $I(x,y)$ is the interference pattern distribution. Holding one beam and introducing a shifted phase to the other, the shifted phase difference of the two beams could lead to the displacement of interference fringes. For phase calibration of the SLM, the interference fringes could be generated by superposing the modulated beam from the SLM with a foregone reference beam. By changing the grayscale of the pattern uploaded on SLM, the modulated phase could be derived from the displacement of the fringes or other changes on the interferograms. In the interferometric methods, the reference beam could come from a certain zone or outside of SLM. The measured phase values are the relatively shifted phase introduced by changing the uploaded grayscale pattern on SLM, instead of the absolute phase of the modulated beam. For the interferometric calibration methods, the beams with approximately equal intensity are needed for good contrast on the interferograms.

3.1. Self-Referenced Method

The self-referenced method is one of the interferometric phase calibration methods which uses different parts of the SLM's active area to realize interference patterns. In this method, the active area in SLM could be divided into two or more zones, to act as a referenced zone and measured phase-shifted zone, respectively. To generate the measured beam, a series of grayscale patterns are uploaded on the modulated zone and the wave passing through the region will be modulated with a shifted phase related to the addressed grayscale on the SLM pixels. Meanwhile, the addressed grayscale on the referenced zone remains unchanged or changed periodically to realize interference with the measured beam. The shifted phase could be derived from the displacement or deformation of the interference fringes.

The earliest experimental arrangement for realizing interference is developed by Young, which inspires the calibration procedure naturally. Nowadays, the most popular phase modulation measurements experimental setup recommended by SLM suppliers in the manual book is based on the division of wavefronts as shown in Figure 3. The wavelength of the light source depends on the working wavelength range of the SLM. A camera is used for capturing the interference fringes. The measurement of the modulated phase could be carried out by remaining the grayscale of one half of the SLM a constant grayscale, while changing the grayscale of the other half of the SLM from 0 to 255 in grayscale gradually. The relatively shifted phase caused by changing the grayscale can be derived from Equation (3) and they could be one to one corresponding to the known grayscales for establishing the grayscale-phase relation. The measured grayscale-phase relation or LUT for the gamma curve could be sent to the driver of SLM by ports according to the manual for better performance of SLM. The setup for the phase measurement of transmissive LC-SLM is shown in reference [13].

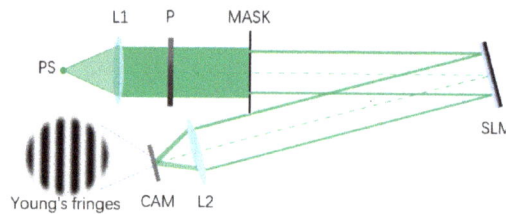

Figure 3. Configuration of the self-referenced phase calibration method for reflective LCoS (HOLOEYE GAEA-2). PS: monochromatic point light source; L1: collimating lens; P: polarizer; MASK: mask with two small holes; L2: Fourier lens; SLM: spatial light modulator; CAM: camera.

In this method, the phase calibration is easy to align for the compact optical structure. However, only the global grayscale-phase relation could be derived. This is because the phase change of the half active area is represented by a beam with the width of a hole on the mask. In addition, the spatial nonuniformity of the SLM could not be characterized by using this method.

A good solution for spatial nonuniformity measurement has been proposed [21,27]. The configuration is shown in Figure 4a. Firstly, the global LUT is measured. The active area on SLM (PLUTO, Holoeye Photonics AG, Germany) is sorted into two zones, the uniform zone and the grating zone. The incident beam hit on both the uniform zone and grating zone. The uniform zone is the uploaded patterns in which the grayscales are uniformly distributed and changed from 0 to 255 at the step of 1. In addition, the grating zone is uploaded with patterns of columns of binary grayscales to form a phase grating. The period of the grating could be chosen according to the optical configuration. The reflected beam from the grating zone could be diffracted into the zeroth order and the first orders. One of the first orders beams interferes with the beam coming from the uniform zone to make the interference fringes. As shown in Figure 4a, the shifted phase of the uniform zone could be derived by the interference fringe's displacement related to the changed grayscales.

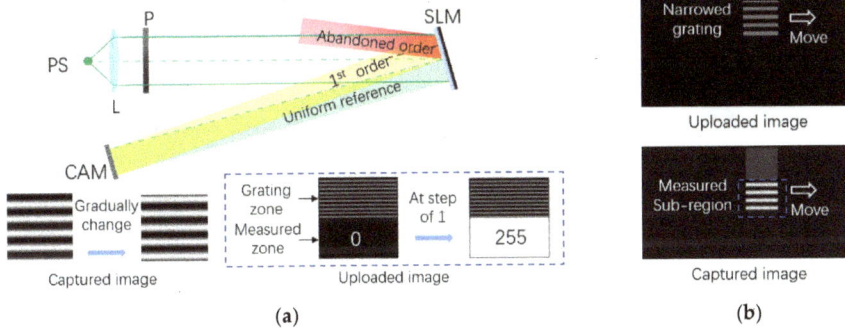

Figure 4. (**a**) Configuration of the self-referenced phase calibration method able to measure spatial nonuniformity. PS: monochromatic point light source; L: collimating lens; P: polarizer; SLM: spatial light modulator; CAM: camera. (**b**) Grayscale pattern uploaded on the SLM for the nonuniformity measurement and image captured by the camera [21].

The fringes move continuously, and the camera used for capturing the fringes is a discrete device, instead. The fringes' displacements smaller than one-pixel size of the camera could be spotted by grayscales changes on the captured image. By using discrete Fourier transform [21], the fringes' displacements could be derived from arguments of two peaks in the Fourier domain, which vary with the fringes' positions or grayscales of interference on the captured image.

This method could be utilized for spatial nonuniformity measurement as well. Before the measurement, a rectangular array of a 128-pixel-diameter pupil mask is uploaded sequentially on SLM for matching the SLM and camera coordinates. During the spatial nonuniformity measurement, the SLM is uploaded with the pattern of a 128-pixel width-narrowed grating unit on a uniform grayscale background as shown in Figure 4b. The grating unit is composed of eight 16-pixel-period columns, and the first diffraction orders diffracted by the narrowed grating will cause interference with the beams coming from the uniform area. By changing the grayscale on the background uniform zone and analyzing the fringes' displacements on the captured image, the shifted phase on the 128 × 128-pixel blocks of uniform background zones could be acquired. The spatial nonuniformity measurement could be carried out by moving the grating unit and testing the shifted phase in the different regions of SLM, and the grayscale-phase relation for sub-regions around the SLM panel (128 × 128 pixels for instance) could be built. To improve the spatial resolution, a smaller grating unit could be built and the spatial resolution of several pixels could be reached. For improving the performance of SLM with calibration, the grayscale image for uploading could be pre-modified by matrix operation for each sub-region [11].

The self-referenced phase calibration method could be simple and easy to apply, which makes it a suggested method in the manual of the SLM products. However, the reference beam comes from a certain active area of the SLM, which means that the measured phase irregular or distortion may come from the reference zones instead of the measured zones. In addition, it could lead to a misunderstanding on the phase measurement of the measured beam. For minimizing the phase distortion from the self-reference beam, a joint analysis of the interferograms for the same region on SLM could be carried out. Considering that the reference beam can be held still, so can the phase distortion induced by the reference beam, the compensate could be carried out.

3.2. Out-Referenced Method

The out-referenced interferometric method uses the reference wave coming from outside the SLM to realize the interference with the measured modulated beam. It is flexible to adjust the foregone reference wave to make a variety of setup transformations, such as the four-step phase-shifting alignment with piezo-mirror which could measure global LUT and the spatial uniformity [28,29].

Another method is using a reference mirror to compose a Twyman-Green interferometer calibration system [30,31]. With all kinds of interferometric setups, the phase nonuniformity of SLM can be measured precisely. Considerations regarding the avoidance of environmental turbulence should be made, such as air fluctuation.

Digital holography technology plays a role in phase recording and reconstruction [32], and it is reasonable to draw lessons from it. One of the out-referenced SLM calibration methods inspired by digital holographic microscopy (DHM) is shown in Figure 5 [33]. The illumination wavelength is 532 nm, and the modulated wavefronts of SLM (Holoeye Pluto, resolution 1920 × 1080, pixel pitch 8 microns) are gathered and magnified by a microscope objective (×10, NA = 0.25). As one microscope objective is used for pixel magnification, the other microscope objective is used for compensation of the quadratic phase term [34]. A camera is used for capturing the hologram of the SLM. There is an offset angle between the reference beam and object beam so that the twin image can be filtered in the spectrum domain [35]. Theoretically, this DHM setup is able to reach the diffraction resolution of approximately 1.6 microns according to the Rayleigh criterion for coherent illumination [36]. Apparently, the single pixel sized phase calibration can be realized using this setup. However, the size of the SLM active area is usually larger than 10 mm × 20 mm, and the FOV(field of view) of the DHM setup is roughly smaller than 10 mm × 10 mm due to a limited space bandwidth [37]. Therefore, the displacements of CCD are needed for capturing the complete SLM active area.

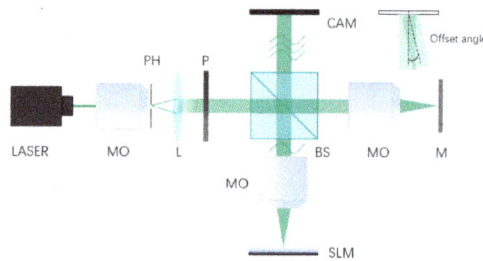

Figure 5. Schematic of the SLM calibration method based on digital holographic microscopy. MO: microscope objective; PH: pinhole filter; L: collimating lens; P: polarizer; BS: beam splitter; M: mirror; SLM: spatial light modulator; CAM: camera. [33].

The methods based on DHM [14] and phase shifting [29] are the two calibration methods able to reach single pixel spatial resolution. However, not all the proposed methods using sophisticated setups to realize the quantitative phase imaging could be used directly for phase calibration, such as Fourier Ptychography [38]. The active area of SLM is usually larger than the FOV of the accurate quantitative phase imaging system with a good horizontal resolution. In addition, there is also the limitation of illumination.

4. Diffractive Phase Calibration Method

The diffractive phase calibration method uses the SLM uploaded with virtual optical elements or coded grayscale patterns to transform the phase distribution into the spectrum or unique intensity features. There are also other methods based on the polarization technique. Compared to the interferometric methods, some of the diffractive calibration systems are more compact with less optical elements, and some distortion induced by the reference beam could be avoided. Besides, some diffractive calibration methods could be carried out without replacing the on-duty setup [39], which makes it suitable for satisfying extremely high-precision requirements.

4.1. Diffraction Pattern Analyzed Method

The diffraction pattern analyzed method is a kind of method using diffraction theory to connect the grayscale-related phase with the directly diffracted pattern changes in the far field. The uploaded grayscale pattern could be a coded image or refreshed simple moving units on a uniform background. The global grayscale-phase relation could be acquired from the diffraction pattern changes, and the spatial nonuniformity could be measured.

In an original diffraction analyzed method [40], an LC television panel removed from Epson projector (VPJ-700) is calibrated. As shown in Figure 6b, a Ronchi grating with two grayscales L_1 and L_2 at 50% duty ratio is designed and the grayscale-related phases of the grating are $\phi_1(L_1)$ and $\phi_2(L_2)$ respectively, so that the relative grayscale and phase are $L = L_1 - L_2$, $\phi(L) = \phi_1(L_1) - \phi_2(L_2)$. The grating period is P, the width of the grating along the y-direction is H, and m is an integer. In addition, the phase transmittance equation along the x-direction can be described as:

$$t(x,y) = \left\{ rect\left(\frac{x-P/4}{P/2}\right) + \exp[i\phi(L)]rect\left(\frac{x-3P/4}{P/2}\right)\right\} \otimes comb(x/P) \cdot rect\left(\frac{x}{mP}\right) \cdot rect\left(\frac{y}{H}\right) \quad (4)$$

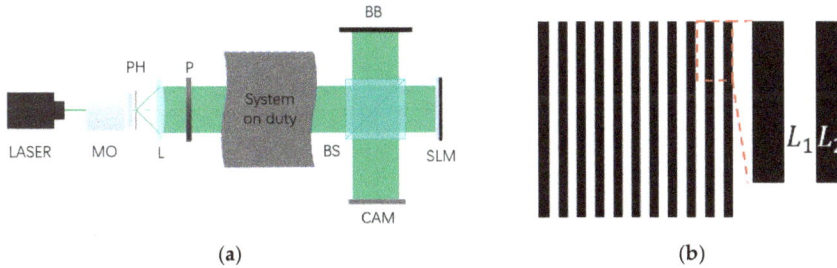

Figure 6. (**a**) Schematic of a diffraction pattern analyzed method. MO: microscope objective; PH: pinhole filter; L: collimating lens; P: polarizer; BS: beam splitter; BB: beam blocker; M: mirror; SLM: spatial light modulator; CAM: camera. (**b**) Uploaded pattern of Ronchi grating.

The *rect* in Equation (4) is the rectangular function, and the *comb* is the comb function. With uniform monochromatic plane wave illumination, spectrum intensity distribution can be described as Fourier transform: $T(f_x, f_y) = F.T.\{t(x,y)\} \cdot 1$. In addition, the spectrum pattern in the Fourier plane along the f_x-axis ($f_y = 0$) is:

$$|T(f_x, 0)|_2 \propto I(f_x) \cdot \cos[\phi(L) + f_x P/2] \quad (5)$$

The relation between the spectrum intensity distribution and the phase difference $\phi(L)$ can be acquired by holding the uploaded L_1 with changing L_2 from 0 to 255 (8-bit) and measuring the intensity changes of the spectrum. A global grayscale-phase relation can be acquired, and this method can also be modified for spatially nonuniformity measurement.

The calibration of SLM suffering from spatially-varying phase response could be realized [11]. The calibrated SLM is from BNS (HSPDM512 1064-PCIe, Boulder Nonlinear Systems). As shown in Figure 7, the pixels on the SLM are composed of 8 × 8-pixel subsections. Because of the small grating size compared to the diffraction distance, the captured image could be analyzed as Fraunhofer diffraction pattern [41]. Neglecting the coefficient, the Fraunhofer diffraction pattern could be treated as the spectrum of the grating. The grayscale-phase relation of the 8 × 8-pixel subsection can be measured according to Equation (5). Hence, the nonuniformity could be measured by moving the small grating inside the subsection.

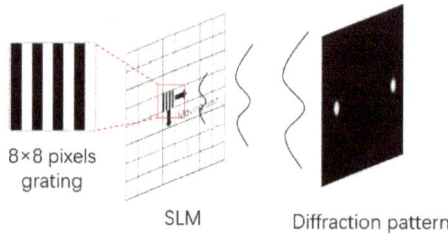

Figure 7. Pattern of single 8 × 8-pixel-grating unit uploaded for spatial nonuniformity measurement.

The checkerboard pattern can also be diffracted for calibration based on the diffractive method [42]. A checkerboard phase pattern with binary phase distribution is uploaded and the far field Fraunhofer diffraction pattern is captured. The related phase could be measured by analyzing the intensity of first orders in the spectrum domain.

By using this kind of diffractive method based on spectrum or Fraunhofer diffraction analysis, the global grayscale-phase LUT and nonuniformity measurement could be carried out. However, a desired camera with a high sensitivity to intensity changes and high signal-to-noise ratio is needed, for the reason that the derived phase is related to the measured intensity directly. Some attempts using the phase retrieval algorithm to reconstruct the phase from the diffraction pattern have been presented [43,44].

4.2. Optical Elements Generated Method

The intensity distribution of the light traveling through the optical elements could be affected by the phase distribution inside the elements. In addition, SLM phase modulation could be measured by generating the virtual elements on the SLM. However, not all of the principles of phase optical elements could be referenced to calibration. The phase distribution of the elements should be binary for variable control.

A simple and robust method for determining the calibration function with binary phase Fresnel lenses is proposed [45]. As shown in Figure 8, a Fresnel lens is designed with binary phase level distribution: ϕ_1 and ϕ_2. The pattern of designed lens is uploaded on the SLM (PLUTO, Holoeye Photonics AG, Germany), which is illuminated by a collimated polarized light. The η is the ratio of the irradiance of the first foci and the illumination intensity on the SLM. It can be described as: $\eta = (2/\pi)^2 \sin^2((\phi_2 - \phi_1)/2)$, and the maximum η can be achieved while the $\phi_2 - \phi_1 = \pi$. For the calibration procedure, the ϕ_1 region can be set to 0 grayscale while the ϕ_2 can be changed from 0 to 255. The shifted phase on ϕ_1 regions is zero, and the shifted phase on ϕ_2 regions changes from zero to the max. As the illumination intensity on the SLM is set to be constant, the irradiance change on the first foci is proportional with η. By measuring the irradiance on the first foci while changing ϕ_2 with the fixed ϕ_1, the grayscale-phase relation can be derived.

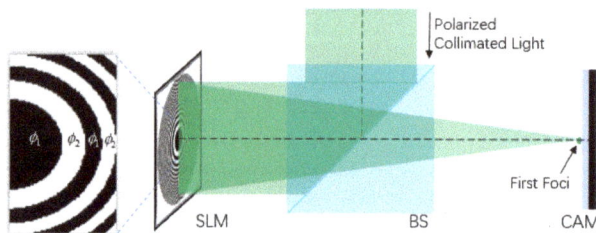

Figure 8. Schematic of an optical elements features analyzed method. BS: beam splitter; SLM: spatial light modulator; CAM: camera.

Using this kind of optical element feature-based calibration method means that a global grayscale-phase LUT could be acquired. Because of the limitation on optical element size for optical feature replication, the spatial nonuniformity measurement is limited.

4.3. Polarization Method

In addition to the radiometric diffraction methods, there are also polarization-based diffraction methods. With the help of the Mueller matrices imaging procedures, the spatial nonuniformity of SLM could be measured with a single pixel resolution (~11.5 µm) [46]. However, this method is very complicated. Another polarization-based system is simple and easy to align, but only the global grayscale-phase relation could be derived [47].

5. Discussion

Two categories of phase calibration methods for SLM are introduced. All of the calibration procedures promote the SLM performance on phase modulation. However, due to the different principles used in the methods, they are different in many ways.

As shown in Table 1, all the calibration methods for SLM could calibrate global LUT and measure spatial nonuniformity, except for the optical elements-generated method. For the spatial nonuniformity measurement of the self-referenced calibration method, the self-referenced beam from certain zones on SLM is supposed to be uniform in theory. But it may not be ideally uniform in practice. Therefore, the spatial nonuniformity measurement may be carried out with errors. As for the single pixel level phase measurement, the out-referenced interferometric method reaches the accuracy of single-pixel resolution with flexible setups. The single pixel measurement ability comes with more optical elements and a complicated setup configuration compared with other calibration methods. The setups of diffraction methods and self-referenced interferometric methods are compact and robust concerning the environmental disturbance. The most charming feature is that the calibration can be carried out without replacing the SLM. In addition, it is practical for optical alignments with high precision. The proposed polarization methods are like a lever with complexity and spatial resolution on both sides. The method which is able to measure the pixel level spatial nonuniformity comes with a complicated data-processing procedure. The simple one could only realize the global measurement.

Table 1. Phase calibration methods with best spatial resolution.

Calibration Method	Resolution (Pixel Pitch)	Frames Needed [1]	Basic Alignment [2]	Reference [3]
Self-referenced	− 8 µm	<20	Directly diffractive	[27]
Out-referenced	1-pixel 8 µm	Several	Twyman-Green	[28,33]
Diffraction pattern analyzed	8 × 8-pixel 14 µm	<10	Directly diffractive	[11]
Optical elements generated	Global	1	Directly diffractive	[45]
Polarization methods	1-pixel 11.5 µm	Several	Retro-reflection	[46,47]

[1] Images or data groups needed approximately for 512 × 512 pixels' calibration at one grayscale. [2] Basic alignments for the best spatial resolution. [3] The parameters are listed according to the discussed referenced paper.

As shown in Figure 9, all the calibration methods mentioned in the article are presented under the consideration of five aspects: system simplicity; spatial resolution; measured region size; environment disturbance isolation; and SLM replacement after calibration. The measured region size is calculated by multiplying the pixel number of spatial resolution by the pixel pitch. The system simplicity,

SLM replacement, and environmental disturbance isolation are evaluated at three levels: marginal, acceptable, and good. In addition, the diffraction pattern analyzed methods could be more functional with considerations of the five aspects.

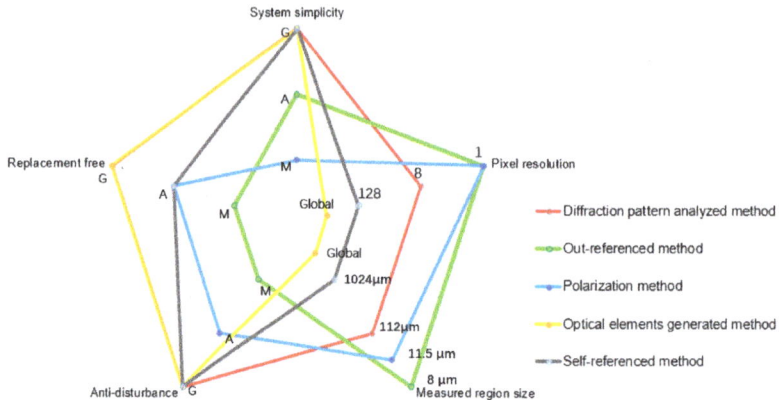

Figure 9. Comparison of features of calibration methods discussed in this review. System simplicity: simplicity of the calibration system; Pixel resolution: achieved spatial resolution in the discussed methods; Measured region size: calculated measured region size of nonuniformity measurement; Anti-disturbance: environment disturbance isolation; Replacement free: flexibility of deploying calibration procedure in an on-duty system without replacing the SLM. M: marginal; A: acceptable; G: good.

6. Conclusions

The SLM phase calibration is essential for realizing high-precision phase modulation. The out-referenced method is suitable for high-precision systems due to its single pixel calibration ability; however, the SLM needs to be replaced. The polarization method could reach a high phase measurement accuracy of approximately 11.5 µm using a polarimeter, but the cost is complicated data processing. The self-referenced methods may come with inherited phase distortion. The diffraction pattern analyzed calibration methods are more practical for in-situ calibration. The spatial nonuniformity measurement could be carried out with easy alignments. With the help of advanced algorithms for phase reconstruction, the diffractive method could be more powerful.

Author Contributions: R.L. made the original draft preparation. R.L. and L.C. made the editing.

Funding: This research is supported by the National Key R & D Program of China (No. 2018YFF0212302), National Key R & D Program of China (No. 2017YFF0106400) and National Natural Science Foundation of China (No. 61827825).

Conflicts of Interest: The authors declare no conflict of interest.

References

1. Zaperty, W.; Kozacki, T.; Kujawińska, M. Multi-SLM color holographic 3D display based on RGB spatial filter. *J. Disp. Technol.* **2016**, *12*, 1724–1731. [CrossRef]
2. Han, Z.; Yan, B.; Qi, Y.; Wang, Y.; Wang, Y. Color holographic display using single chip LCOS. *Appl. Opt.* **2019**, *58*, 69–75. [CrossRef]
3. Koba, M.I. High-Resolution Spatial Light Modulator for 3-Dimensional Holographic Display. U.S. Patent 6819469B1, 16 November 2004.
4. Reynoso-Alvarez, A.; Strojnik, M.; García-Torales, G.; Flores-Nuñez, J.L. A high precision phase-shifter modulator in a shearing interferometric system. In Proceedings of the Infrared Remote Sensing and Instrumentation XXV, San Diego, CA, USA, 30 August 2017.

5. Kacperski, J.; Kujawinska, M. Phase only SLM as a reference element in Twyman-Green laser interferometer for MEMS measurement. In Proceedings of the Optical Measurement Systems for Industrial Inspection V, Munich, Germany, 18 June 2007.

6. Wulff, K.D.; Cole, D.G.; Clark, L.R.; Dileonardo, R.; Leach, J.; Cooper, J.; Gibson, G.; Padgett, M.J. Aberration correction in holographic optical tweezers. *Opt. Express* **2006**, *14*, 4169–4174. [CrossRef] [PubMed]

7. Peng, T.; Li, R.; An, S.; Yu, X.; Zhou, M.; Bai, C.; Liang, Y.; Lei, M.; Zhang, C.; Yao, B.; et al. Real-time optical manipulation of particles through turbid media. *Opt. Express* **2019**, *27*, 4858–4866. [CrossRef]

8. Duerr, P.; Dauderstaedt, U.; Kunze, D.; Auvert, M.; Bakke, T.; Schenk, H.; Lakner, H. Characterization of spatial light modulators for microlithography. In Proceedings of the MOEMS Display and Imaging Systems, San Jose, CA, USA, 20 January 2003.

9. Tukker, T.W.; Pandey, N.; Verschuren, C.A. Illumination System for a Lithographic or Inspection Apparatus. U.S. Patent 20180004095A1, 1 April 2018.

10. Mikhaylov, D.; Kiedrowski, T.; Lasagni, A.F. *Beam Shaping Using Two Spatial Light Modulators for Ultrashort Pulse Laser Ablation of Metals*; SPIE: San Francisco, CA, USA, 2019; Volume 10906.

11. Engström, D.; Persson, M.; Bengtsson, J.; Goksör, M. Calibration of spatial light modulators suffering from spatially varying phase response. *Opt. Express* **2013**, *21*, 16086–16103. [CrossRef]

12. Bentley, J.B.; Davis, J.A.; Albero, J.; Moreno, I. Self-interferometric technique for visualization of phase patterns encoded onto a liquid-crystal display. *Appl. Opt.* **2006**, *45*, 7791–7794. [CrossRef]

13. Bergeron, A.; Gauvin, J.; Gagnon, F.; Gingras, D.; Arsenault, H.H.; Doucet, M. Phase calibration and applications of a liquid-crystal spatial light modulator. *Appl. Opt.* **1995**, *34*, 5133–5139. [CrossRef]

14. Robertson, B.; Zhang, Z.; Redmond, M.M.; Collings, N.; Liu, J.; Lin, R.S.; Jeziorska-Chapman, A.M.; Moore, J.R.; Crossland, W.A.; Chu, D.P. Use of wavefront encoding in optical interconnects and fiber switches for cross talk mitigation. *Appl. Opt.* **2012**, *51*, 659–668. [CrossRef]

15. Zhang, Z.; You, Z.; Chu, D. Fundamentals of phase-only liquid crystal on silicon (LCOS) devices. *Light Sci. Amp. Appl.* **2014**, *3*, e213. [CrossRef]

16. Raynes, E.P. Optically active additives in twisted nematic devices. *Rev. Phys. Appliq.* **1975**, *10*, 117–120. [CrossRef]

17. Schadt, M.; Helfrich, W. Voltage-dependent optical activity of a twisted nematic liquid crystal. *Appl. Phys. Lett.* **1971**, *18*, 127–128. [CrossRef]

18. Yeh, P.; Gu, C. *Optics of Liquid Crystal Displays*, 2nd ed.; Wiley: Hoboken, NJ, USA, 2010; pp. 792–793.

19. Dasgupta, P.; Das, M.K.; Das, B. physical properties of three liquid crystals with negative dielectric anisotropy from X-ray diffraction and optical birefringence measurements. *Mol. Cryst. Liquid Cryst.* **2011**, *540*, 154–161. [CrossRef]

20. Kohns, P.; Schirmer, J.; Muravski, A.A.; Yakovenko, S.Y.; Bezborodov, V.; Dābrowsk, R. Birefringence measurements of liquid crystals and an application: An achromatic waveplate. *Liquid Cryst.* **1996**, *21*, 841–846. [CrossRef]

21. Fuentes, J.L.M.; Fernánde, E.J. Interferometric method for phase calibration in liquid crystal spatial light modulators using a self-generated diffraction-grating. *Opt. Express* **2016**, *24*, 14159–14171. [CrossRef] [PubMed]

22. Yokoyama, H.; van Sprang, H.A. A novel method for determining the anchoring energy function at a nematic liquid crystal-wall interface from director distortions at high fields. *J. Appl. Phys.* **1985**, *57*, 4520–4526. [CrossRef]

23. Zaperty, W.; Kozacki, T.; Gierwiało, R.; Kujawińska, M. The RGB imaging volumes alignment method for color holographic displays. In Proceedings of the Photonics Applications in Astronomy, Communications, Industry, and High-Energy Physics Experiments, Wilga, Poland, 28 September 2016.

24. Zhang, H.; Xie, J.; Liu, J.; Wang, Y. Elimination of a zero-order beam induced by a pixelated spatial light modulator for holographic projection. *Appl. Opt.* **2009**, *48*, 5834–5841. [CrossRef] [PubMed]

25. Persson, M.; Engström, D.; Goksör, M. Reducing the effect of pixel crosstalk in phase only spatial light modulators. *Opt. Express* **2012**, *20*, 22334–22343. [CrossRef] [PubMed]

26. Martínez, J.L.; Moreno, I.; del Mar Sánchez-López, M.; Vargas, A.; García-Martínez, P. Analysis of multiple internal reflections in a parallel aligned liquid crystal on silicon SLM. *Opt Express* **2014**, *22*, 25866–25879.

27. Zhao, Z.; Xiao, Z.; Zhuang, Y.; Zhang, H.; Zhao, H. An interferometric method for local phase modulation calibration of LC-SLM using self-generated phase grating. *Rev. Sci. Ins.* **2018**, *89*, 083116. [CrossRef]

28. Mukhopadhyay, S.; Sarkar, S.; Bhattacharya, K.; Hazra, L. *Polarization Phase Shifting Interferometric Technique for Phase Calibration of a Reflective Phase Spatial Light Modulator*; SPIE: Bellingham, WA, USA, 2013; Volume 52, pp. 1–7.

29. Xun, X.; Cohn, R.W. Phase calibration of spatially nonuniform spatial light modulators. *Appl. Opt.* **2004**, *43*, 6400–6406. [CrossRef]

30. Zhang, H.; Zhang, J.; Wu, L. Evaluation of phase-only liquid crystal spatial light modulator for phase modulation performance using a Twyman–Green interferometer. *Meas. Sci. Technol.* **2007**, *18*, 1724–1728. [CrossRef]

31. Otón, J.; Ambs, P.; Millán, M.S.; Pérez-Cabré, E. Multipoint phase calibration for improved compensation of inherent wavefront distortion in parallel aligned liquid crystal on silicon displays. *Appl. Opt.* **2007**, *46*, 5667–5679. [CrossRef]

32. Schnars, U.; Falldorf, C.; Watson, J.; Jüptner, W. (Eds.) *Digital Holography*; Springer: Berlin, Germany, 2015; Chapter II; pp. 39–68.

33. Yang, L.; Xia, J.; Chang, C.; Zhang, X.; Yang, Z.; Chen, J. Nonlinear dynamic phase response calibration by digital holographic microscopy. *Appl. Opt.* **2015**, *54*, 7799–7806. [CrossRef]

34. Xia, J.; Zhu, W.; Heynderickx, I. 41.1: Three-dimensional Electro-Holographic Retinal Display. In *SID Symposium Digest of Technical Papers*; Blackwell Publishing Ltd.: Oxford, UK, 2011; pp. 591–594.

35. Cuche, E.; Marquet, P.; Depeursinge, C. Spatial filtering for zero-order and twin-image elimination in digital off-axis holography. *Appl. Opt.* **2000**, *39*, 4070–4075. [CrossRef]

36. Cotte, Y.; Toy, F.M.; Jourdain, P.; Pavillon, N.; Boss, D.E.; Magistretti, P.J.; Marquet, P.; Depeursinge, C.D. Marker-free phase nanoscopy. *Nat. Photon.* **2013**, *7*, 113. [CrossRef]

37. Lohmann, A.W.; Dorsch, R.G.; Mendlovic, D.; Zalevsky, Z.; Ferreira, C. Space–bandwidth product of optical signals and systems. *J. Opt. Soc. Am. A* **1996**, *13*, 470–473. [CrossRef]

38. Lee, B.; Hong, J.-Y.; Yoo, D.; Cho, J.; Jeong, Y.; Moon, S.; Lee, B. Single-shot phase retrieval via Fourier ptychographic microscopy. *Optica* **2018**, *5*, 976–983. [CrossRef]

39. Engström, D.; Persson, M.; Goksör, M. Spatial Phase calibration used to improve holographic optical trapping. In Proceedings of the Biomedical Optics and 3-D Imaging, Miami, FL, USA, 28 April 2012.

40. Zhang, Z.; Lu, G.; Francis, T.; Yu, S. *Simple Method for Measuring Phase Modulation in Liquid Crystal Televisions*; SPIE: Bellingham, WA, USA, 1994; Volume 33, pp. 3018–3022.

41. Born, M.; Wolf, E. *Principles of Optics*, 7th ed.; Cambridge University Press: Cambridge, UK, 1999; pp. 425–430, Chapter VIII.

42. Chen, X.; Chen, X.; Li, J.; Chen, D. A calibration algorithm for the voltage-phase characteristic of a liquid crystal optical phased array. In Proceedings of the Selected Papers of the Photoelectronic Technology Committee Conferences, Harbin, China, 5 November 2015.

43. Kohler, C.; Zhang, F.; Osten, W. Characterization of a spatial light modulator and its application in phase retrieval. *Appl. Opt.* **2009**, *48*, 4003–4008. [CrossRef]

44. Hart, N.W.; Roggemann, M.C.; Sergeyev, A.V.; Schulz, T.J. *Characterizing Static Aberrations in Liquid Crystal Spatial Light Modulators Using Phase Retrieval*; SPIE: Bellingham, WA, USA, 2007; Volume 46, pp. 1–7.

45. Mendoza-Yero, O.; Mínguez-Vega, G.; Martínez-León, L.; Carbonell-Leal, M.; Fernández-Alonso, M.; Doñate-Buendía, C.; Pérez-Vizcaíno, J.; Lancis, J. Diffraction-based phase calibration of spatial light modulators with binary phase fresnel lenses. *J. Disp. Technol.* **2016**, *12*, 1027–1032. [CrossRef]

46. Wolfe, J.E.; Chipman, R.A. Polarimetric characterization of liquid-crystal-on-silicon panels. *Appl. Opt.* **2006**, *45*, 1688–1703. [CrossRef]

47. Martínez, F.J.; Márquez, A.; Gallego, S.; Ortuño, M.; Francés, J.; Beléndez, A.; Pascual, I. Averaged stokes polarimetry applied to evaluate retardance and flicker in PA-LCoS devices. *Opt. Express* **2014**, *22*, 15064–15074. [CrossRef] [PubMed]

applied
sciences

MDPI

Review

Helix-Free Ferroelectric Liquid Crystals: Electro Optics and Possible Applications

Alexander Andreev, Tatiana Andreeva, Igor Kompanets * and Nikolay Zalyapin

P. N. Lebedev Physical Institute, Moscow 119991, Russia; ALA-2012@yandex.ru (A.A.);
rybusenok@yandex.ru (T.A.); nikolay.zal@gmail.com (N.Z.)
* Correspondence: kompan@sci.lebedev.ru; Tel.: +7-903-124-3235

Received: 7 November 2018; Accepted: 25 November 2018; Published: 29 November 2018

Featured Application: As the materials for fast low-voltage displays and light modulators.

Abstract: This is a review of results from studying ferroelectric liquid crystals (FLCs) of a new type developed for fast low-voltage displays and light modulators. These materials are helix-free FLCs, which are characterized by spatially periodic deformation of smectic layers and a small value of spontaneous polarization (less than 50 nC/cm^2). The FLC director is reoriented due to the motion of solitons at the transition to the Maxwellian mechanism of energy dissipation. A theoretical model is proposed for describing the FLC deformation and director reorientation. The frequency and field dependences of the optical response time are studied experimentally for modulation of light transmission, scattering, and phase delay with a high rate. The hysteresis-free nature and smooth dependence of the optical response on the external electric field in the frequency range up to 6 kHz is demonstrated, as well as bistable light scattering with memorization of an optical state for a time exceeding the switching time by up to 6 orders of magnitude. Due to the spatially inhomogeneous light phase delay, the ability of a laser beam to cause interference is effectively suppressed. The fastest FLCs under study are compatible with 3D, FLC on Silicon (FLCoS), and Field Sequential Colors (FSC) technologies.

Keywords: liquid crystal; ferroelectric; helix-free; soliton; transparent mode; light scattering; speckle suppression

1. Introduction

It is known that the minimum-time optical response is achieved in some smectic liquid crystals, called smectics C*, which possess ferroelectric properties and high sensitivity to electric fields [1–3]. The principle of electro-optical modulation in ferroelectric liquid crystals (FLCs), like in widely used nematic liquid crystals (NLCs), is the electrically controlled birefringence or light scattering.

A distinctive feature of smectic crystals is the layered structure formed as a result of ordering the centers of mass of FLC molecules along the direction of orientation of their long axis (director), with a pitch of the order of the molecules' length. Among the smectic crystals, the helix FLCs are the most well known, in which the polar axes of various smectic layers are rotated relative to each other, forming a helix (spiral) twist of the FLC director in the absence of an external electric field (Figure 1). In each layer, the position of the director **n** is determined by the polar angle Θ_0 and the azimuth angle φ, which varies from 0 to 2π at a distance equal to the pitch p_0 of the helix. Under the action of the electric field **E**, which is parallel to smectic layers (along the coordinate x), the vector **P**$_S$ of spontaneous polarization is oriented in all layers along the field direction. As a result, the director acquires one direction in the entire volume of the FLC layer—the direction of the FLC main optical axis.

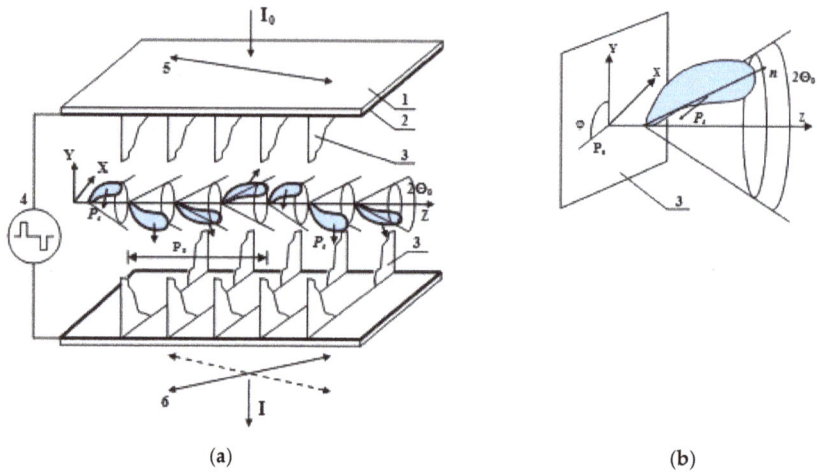

Figure 1. The helix ferroelectric liquid crystal (FLC)-based electro-optical cell with planar orientation of a layer (**a**) and mutual location of the spontaneous polarization vector P_S, a smectic layer, and FLC director n (**b**). I_0 and I are the intensities of the incident light and the light passing through an FLC cell, correspondingly. 1—glass substrates with conductive covers 2; 3—smectic layers; 4—generator of bipolar pulses; 5 and 6—polarizers; p_0—helix pitch; Θ_0—angle of molecule tilt in smectic layers.

When a sign of the field E changes, the orientation of the vector P_S changes by 180°, and the long axes of molecules unfold along the cone with the generatrix $2\Theta_0$, resulting in a change of the angle φ by 180°. The reorientation of the director determining the main optical axis of the ellipsoid of the FLC refractive indices results in a change in the angle between the polarization plane of the incident light (I_0) and the main optical axis of the ellipsoid. This leads to the modulation of the phase delay between the ordinary and extraordinary rays, or to light intensity modulation, if the electro-optic cell is placed between crossed polarizers [1–3]. Under certain conditions, electrically controlled light scattering on so-called transient domains is observed in the helix FLC due to the formation of a spatially inhomogeneous structure of refractive index gradients in the FLC layer [4,5].

Helix-free (no spiral) FLCs are also known [6]. In them, the helicoidal twist of the director in the FLC volume is compensated or suppressed by the interaction of chiral optically active additives with opposite signs of optical activity. The coincidence of the same signs of spontaneous polarization in chiral additives makes it possible to obtain a value of spontaneous polarization of 100 nC/cm² or higher for FLCs with a compensated helicoid.

In contrast to NLC, the electro-optical effect in FLC is linear relative to the field [7,8], and since FLC reacts to the sign of the applied electric voltage, the values of the on and off times of the optical response are the same and are proportional to

$$\tau_R \sim \gamma_\varphi / P_S \times E, \tag{1}$$

if the dissipative coefficient is the rotational viscosity γ_φ. The FLC is returned to its original state by a reverse-polarity pulse, i.e., forcedly, not as a result of relaxation due to elastic forces (like in the NLC). Therefore, the optical response at switching on and off is symmetrical and very short (in the sub-millisecond region), especially for FLC with low viscosity and large spontaneous polarization. However, it is impossible to significantly reduce τ_R by increasing the spontaneous polarization, since this usually leads to an increase in the FLC rotational viscosity. This circumstance limits the frequency of light modulation if the applied voltage does not exceed several volts.

In the research carried out at the Lebedev Physical Institute (LPI), it was shown that the nature of the FLC director reorientation in an electric field depends essentially on the coefficient responsible for the energy dissipation in an FLC layer: the rotational or shear viscosity [9]. If the alternating electric field of the frequency f acts on the FLC, and the period of its variation is large ($\tau_m \cdot f << 1$) in comparison with the Maxwellian relaxation time τ_m [10], then the FLC behaves as a liquid with the viscosity γ_φ. On the contrary, at sufficiently high frequencies ($\tau_m \cdot f >> 1$), the FLC behaves as an amorphous solid, and the dissipative coefficient is the shear viscosity (denoted γ_ψ).

The predominance of shear viscosity leads to a change in the character of the motion of the helix FLC director in weak electric fields: the reorientation occurs due to the motion of 180°-domain walls [9,11,12]. Such a reorientation process made it possible to obtain a modulation frequency of light radiation of the order of 3 kHz with an electric field strength of about 1 V/μm. The time of the electro-optical response was $50 \div 70$ μs. As a disadvantage, some distortion of the spectral composition of the modulated radiation and the residual light scattering caused by the presence of a helix was noted.

The character of the electric field action is practically the same for known helix-free FLCs. In them, the azimuth angle φ of the director's orientation in the volume of a layer is a value that is practically constant in all smectic layers; i.e., there is a so-called spatially homogeneous structure. The magnitude of the spontaneous polarization is much higher than 50 nC/cm^2. The dissipative coefficient is the rotational viscosity γ_φ, and the director reorientation time, as for the helix FLC, does not depend on the frequency of the electric field change [8].

A new type of helix-free FLC specially developed at the LPI has a different character of interaction with the electric field. This material is a spatially inhomogeneous structure with periodic deformation of smectic layers (with a pitch from 1.5 to 5 μm) in the absence of an electric field and a relatively small value of spontaneous polarization (less than 50 nC/cm^2). In such an FLC, the director reorientation in the alternating electric field is due to the motion of structurally stable localized waves of a stationary profile—dynamic solitons that arise upon the transition to the Maxwellian mechanism of energy dissipation—and the electro-optical response depends essentially on the frequency of the electric field change [13–16]. Note that a director reorientation through solitons was proposed for the first time in [17].

The new materials provided unique parameters of radiation modulation unattainable for LC analogs. For example, in a transparent mode, in an electric field of the order of 1 V/μm (at a control voltage of ±1.5 V), experimental samples of electro-optical cells with the novel helix-free FLC show a modulation characteristic with the fastest optical response (about 25 microseconds) and the highest modulation frequency (up to 7 kHz), including hysteresis-free characteristics with a continuous gray scale up to 6 kHz. In a bistable light-scattering mode, a state with intensive scattering can be turned on and off for a few tens of microseconds and be memorized for several tens of seconds or until a pulse of opposite polarity appears [16].

We will now consider in more detail the physical properties, light modulation characteristics, and possible applications of the new helix-free FLC based on the results attained from our research during the last few years.

2. Ferroelectric Liquid Crystals Compositions Used and Their Basic Properties

The basis of the new FLCs is the same optically active additives used in known FLC and based on derivatives of terphenyl-dicarboxylic acid. The difference is only in the specific compositions of FLCs, which is the object of know-how. In spite of the great difference in the rotational viscosity coefficient γ_φ, which in various compositions changes by practically an order of magnitude (from 0.15 to 1.5 P), and in spite of the manifestation of essentially different optical properties (to be discussed below), the values of the spontaneous polarization P_S, the initial tilt angle of molecules in smectic layers Θ_0, and the sequence of phase transitions for all new FLCs differ insignificantly (see Table 1).

The sequence of phase transitions for these compositions is the following:

HF-32B: Cr 2 °C → Sm C* 70 °C → Sm A* 101 °C → I,
HF-32C: Cr 1 °C → Sm C* 75 °C → Sm A* 102 °C → I,
HF-32D: Cr 0 °C → Sm C* 68 °C → Sm A* 98 °C → I,
HF-32E: Cr 2 °C → Sm C* 75 °C → Sm A* 103 °C → I,
HF-32F: Cr 1 °C → Sm C* 73 °C → Sm A* 90 °C → I.

Here, *Cr* is the crystalline phase, *Sm C** is the chiral smectic C phase (ferroelectric), *Sm A** is the chiral smectic *A* phase (paraelectric), and *I* is the isotropic (liquid) phase.

Table 1. Investigated helix-free ferroelectric liquid crystals (FLC) compositions.

Compositions	Θ_0, grad.	P_S, nC/cm^2	γ_φ, P
HF-32B	21.7	40	0.7
HF-32C	22	40	1.0
HF-32D	23	42	0.15
HF-32E	22	45	1.5
HF-32F	21	42	0.75

The new helix-free FLC materials exhibit the following basic properties in electro-optical cells, first discovered and investigated at the LPI:

- A periodic spatial deformation of FLC smectic layers with a pitch of 1.5÷6 μm is observed in the absence of an electric field;
- The FLC director is reoriented as a result of the soliton waves motion;
- Certain FLC compositions show the fastest (among all LCs) optical response in a transparent mode with continuous gray scale;
- Certain FLC compositions show the fastest (among all LCs) optical response in the light-scattering mode with data storage;
- Certain FLC compositions provide a rapid change in the phase delay of the modulated radiation initiated by light scattering.

3. Periodic Deformation of Smectic Layers

The periodic deformation of smectic layers in new helix-free FLCs in the absence of an external electric field is the result of compensation of the space charge of the spontaneous polarization. In the case of homeotropic orientation of FLC molecules, where smectic layers are parallel to the electro-optical cell substrates, the periodic deformations are clearly observed (Figure 2) and look behind crossed polarizers like alternating bands with a pitch of 1.5÷6 μm, depending on the FLC molecular structure [15,16].

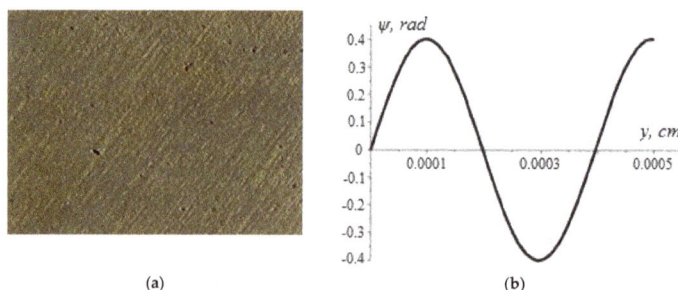

(a) (b)

Figure 2. Alternation of light and dark bands, illustrating the spatially periodic deformation of smectic layers: (**a**) observed experimentally and (**b**) calculated (see Section 5).

A deformation occurs at a certain ratio between the values of FLC essential parameters, as follows:

- The rotational viscosity is in the range of $0.3 < \gamma_\varphi < 1.0$ P. If it is less, the transition to the shear viscosity γ_ψ is not achieved, and the soliton mechanism of reorientation of the FLC director is not realized; if it is more than 1 P, the optical response time increases significantly at higher frequencies, and the transition to the soliton mode is not observed;
- The magnitude of the spontaneous polarization P_S is less than 50 nC/cm^2. If it is greater, the saturation voltage increases, the ferroelectric domains begin to form, and residual light scattering takes place after the electric field is turned off.

The periodical deformation of smectic layers is responsible for the properties of the novel FLCs indicated above. In fact, the fast electro-optical response with continuous gray scale in a transparent mode, which could be used in display devices, is observed in the FLC compositions with a rather long deformation pitch of more than 3 μm. For example, for the composition HF32F, the deformation pitch is of 4 μm. Intensive light scattering in FLC, which could be used in polarization-free devices, manifests in compositions with rather short deformation pitch of $1.5 \div 2.0$ μm (this value for HF32 is about 1.5 μm). A fast-changing phase delay (initiated by light scattering switching on), which could be used for suppressing speckles in laser images, is realized more preferably in FLC compositions with a deformation period of about 3 μm. Although the periods of spatial deformation for HF32F and HF32B are different, other essential parameters of these compositions are almost the same (see Table 1).

4. Ferroelectric Liquid Crystals Director Reorientation by an Electric Field in the Soliton Mode

As shown in [15,16], the periodic deformation of smectic layers means that for FLC molecules initially inclined at the angle Θ_0 with respect to the normal to the FLC layer at a given point, it is energetically preferable to additionally deflect by some angle $\pm\Psi$ with respect to the direction z of substrate rubbing (Figure 3). Because of this, the position of the FLC main optical axis along the smectic layers changes, and the birefringence depends on the electric field frequency.

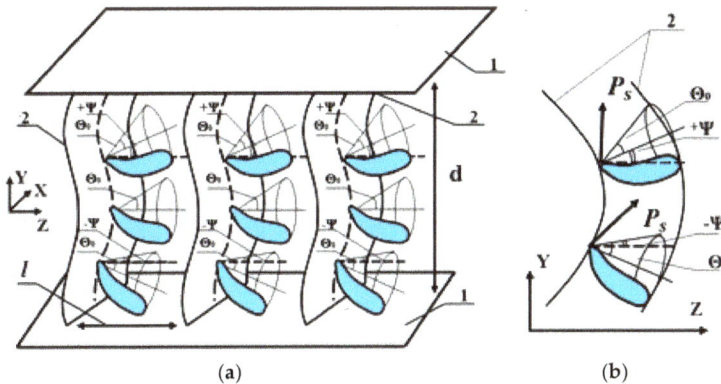

(a)   (b)

Figure 3. An electro-optical cell with helix-free FLC (**a**) and a fragment of a deformed smectic layer (**b**) 1—glass substrates with conrductive covers; 2—smectic layers; Θ_0—the angle of molecule tilt in smectic layers; ψ—angle of tilt of a smectic layer; P_S—vector of spontaneous polarization; d—FLC cell thickness; l—smectic layer thickness.

The electric field E applied along the coordinate y interacts with the spontaneous polarization and changes the director distribution (the angle Ψ) in each smectic layer. The reorientation of the FLC director can occur both with the change of the azimuthal angle φ of the director orientation by 180°, when the director is reoriented along the cone generatrix with its $2\Theta_0$ turn, and with the change in the distribution of the angle ψ characterizing the deformation of smectic layers. In the first case, the dissipative coefficient is the rotational viscosity γ_φ, and in the second case, it is the viscosity for the shear deformation γ_ψ.

When after the electric field is turned off the director relaxation time begins to depend on the electric field frequency, the transition to the Maxwellian mechanism of energy dissipation takes place—that is, the transition from the rotation viscosity γ_φ to the shear viscosity γ_ψ.

To describe the nonlinear dynamic process of FLC director reorientation in an external electric field, we used the soliton approach [15–18] by which many nonlinear effects and processes have been described before, including those in ferroelectric liquid crystals [17,19–21].

The soliton mode manifests in the new FLC compositions possessing coefficients of rotational viscosity in the range from 0.15 to 1 P, when the electric field frequency or, simultaneously, the frequency and the strength increase. At frequencies corresponding to the transition to the Maxwellian mechanism of energy dissipation, the viscosity γ_ψ for the shear strain becomes a dissipative coefficient, and the dynamic solitons move along the smectic layers. The director of the FLC is reoriented by the motion of soliton waves.

The appearance of the soliton waves is due to the joint influence of the medium nonlinearity and the presence of a dispersion of velocities of deformation waves (or displacements). They sharply increase the steepness of the wave front and cause the growth of gradients of the field variables. This leads to a spatial redistribution of the excitation energy and its localization (the wave period tends to infinity). The shape of dynamic solitons is uniquely related to the independent parameters—in particular, to the velocity of the soliton center [18]:

$$V = \frac{\Theta_0}{\gamma_\psi}\left(2K(P_S E \cos\varphi_0 + M) - \left(\frac{2K}{d\Theta_0}\right)^2\right)^{1/2},$$

(2)

where K is the FLC elastic modulus describing the deformation of the director with respect to the angle Ψ; γ_ψ is the shear viscosity; M is the bending energy of smectic layers; and φ_0 is the initial azimuth angle of the director orientation.

The transition to the Maxwellian mechanism of energy dissipation and the soliton mechanism of director reorientation is accompanied by a strong frequency dependence of the electro-optical response time $\tau_{0.1-0.9}$. It can be seen from Figure 4 that in the frequency range from 100 to 200 Hz, this time for the composition HF-32B with $\gamma_\varphi = 0.7$ P decreases by almost half.

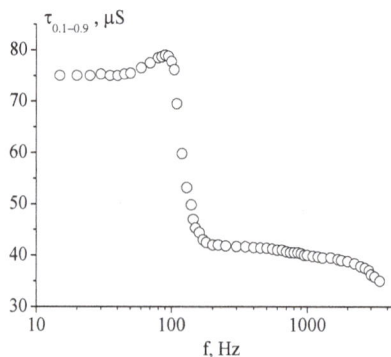

Figure 4. Frequency dependence of the optical response time for the HF-32B composition at V = ±1.5 V. The bipolar voltage is of the rectangular shape (meander). The thickness of the electro-optical cell is 1.7 µm.

For more viscous FLC compositions, for example, for HF-32E with $\gamma_\varphi = 1.5$ Poise, the frequency dependence of the electro-optical response time $\tau_{0.1-0.9}$ is significantly weakened (Figure 5a). The reason for this weakening is that the relaxation time of the FLC director to the unperturbed state after the voltage turning off is practically independent of the frequency of the electric field change (Figure 5b). Consequently, the transition to the Maxwellian mechanism of energy dissipation does not occur, and the soliton mode does not manifest.

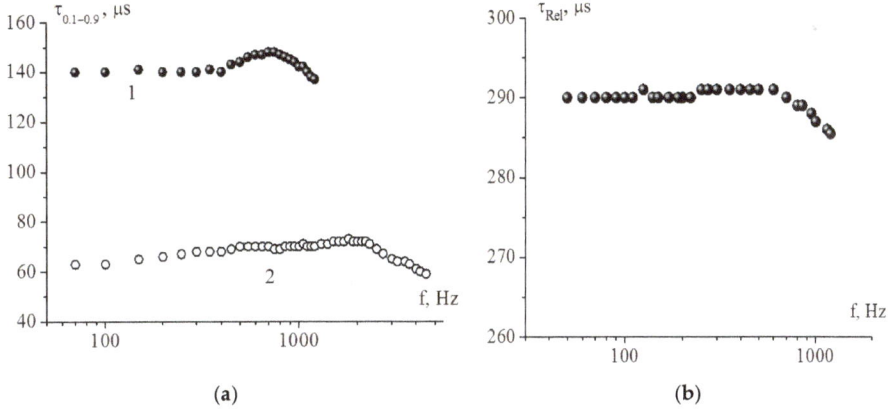

(a) (b)

Figure 5. Frequency dependences (**a**) of the optical response time at ±1.5 V (curve 1) and ±12 V (curve 2) pulses acting and (**b**) of the director relaxation time after ±1.5 V pulses turning off. The FLC composition is HF32E (γ_φ = 1.5 P). A bipolar voltage of rectangular shape (meander) was applied. The thickness of the electro-optical cell is 1.7 µm.

A decrease in the rotational viscosity coefficient to 0.15 Poise (for the composition HF-32D with practically the same spontaneous polarization value) leads to the soliton mode not being observed at the control voltage amplitude ±1.5 V (Figure 6, curve 1). The soliton mode appears when the electric field strength exceeds 5 V/µm (Figure 6, curve 2). This is especially seen in the frequency range of a few hundred Hz, when the transition to the mechanism of energy dissipation takes place.

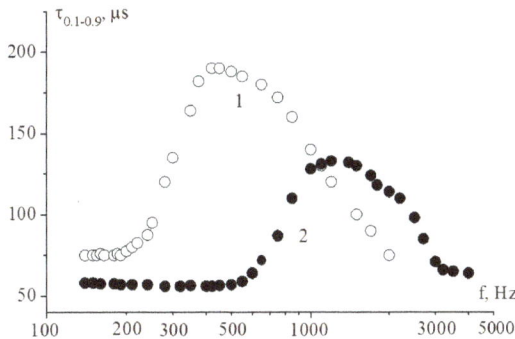

Figure 6. Frequency dependencies of the electro-optical response time for the HF32D composition (γ_φ = 0.15 Poise). 1—amplitude of the bipolar voltage (meander) ±1.5 V; 2—amplitude ±9 V. The thickness of the electro-optical cell is 1.7 µm.

5. Theoretical Model of the Ferroelectric Liquid Crystal Deformation and Director Reorientation

5.1. Deformation of Ferroelectric Liquid Crystal Smectic Layers

The bulk density of free energy associated with the periodic deformation of smectic layers can be written as [16,18,22]

$$F = \frac{1}{2}K\left(\frac{d\psi}{dy}\right)^2 + \frac{1}{2}M\left(\frac{l-l_0}{l_0}\right)^2. \tag{3}$$

Here, K is the elasticity coefficient describing the director's deformation over the angle ψ (Figure 3) and M is the bending energy of smectic layers. We also take into account the smallness of the angles ψ and Θ_0 (ψ, $\Theta_0 \ll 1$), and the relative change of the smectic layer thickness $(l - l_0)/l_0 \cong (\psi^2 - \Theta_0^2)/2$ [23].

If to minimize Equation (3) we can find the distribution of the angle ψ along the coordinate y ($0 \leq y \leq d$, where d is the thickness of the electro-optical cell), then

$$\frac{d^2\psi}{dy'^2} + \frac{M\Theta_0^2 d^2}{2K}\psi - \frac{Md^2}{2K}\psi^3 = 0, \tag{4}$$

where $y' = y/d$ and $0 \leq y' \leq 1$.

The exact solution of Equation (4) is written in terms of the Jacobi elliptic sine:

$$\psi = \sqrt{2}\Theta_0 \frac{k}{\sqrt{1+k^2}} sn\left(\Theta_0 d\sqrt{\frac{M}{2K}}\frac{(y'+C_1)}{\sqrt{1+k^2}}, k\right) \tag{5}$$

where $k = \sqrt{\frac{M\Theta_0^4 d^2}{8KC_2}} - \sqrt{\frac{M\Theta_0^4 d^2}{8KC_2} - 1}$, $0 < C_2 < \frac{M\Theta_0^4 d^2}{8K}$, and $0 < k < 1$ is the modulus of Jacobi elliptic sine.

Using the Van der Pol method [24] and substituting the boundary conditions $\psi(y = 0) = 0$ and $d\psi/dy = 0$ at $\psi = \Theta_0$, we obtain an approximate solution to Equation (4):

$$\psi = \Theta_0 \sin\left(\frac{5}{8}\Theta_0\sqrt{\frac{M}{2K}}y\right). \tag{6}$$

This analytical solution given by Equation (6) describes the structure of smectic layers deformed periodically in the direction y. Maxima and minima in the dependence $\psi(y)$ correspond to light and dark stripes in Figure 2a for a cell with FLC layer thickness of 5 µm, $\Theta_0 = 23°$, $M = 4 \times 10^3$ erg/cm^3, and $K = 5 \times 10^{-12}$ N.

Reorientation of the FLC director due to the interaction of the alternating electric field E applied along the coordinate y (Figure 2b) with the spontaneous polarization P_S can occur both by changing the azimuth angle φ of the director orientation by 180° (if the director is reoriented along the generating lines of a cone with \ apex angle $2\Theta_0$) and by changing the distribution of ψ (deformation of smectic layers). In the first case, the dissipative coefficient is the rotational viscosity γ_φ, and in the second one, the viscosity γ_ψ of the shear deformation predominates [16,18].

5.2. Ferroelectric Liquid Crystal Director Reorientation

The bulk density of the free energy for the electrostatic interaction of the electric field and spontaneous polarization can be written as

$$\frac{1}{2}P_S E \cos\varphi_0\left(\frac{l-l_0}{l_0}\right)^2, \tag{7}$$

where φ_0 is the initial azimuth angle of the director orientation. Now we minimize the expression in Equation (3) and record the equation of a balance of moments, describing the change ψ in the electric field E:

$$-\gamma_\psi \frac{\partial\psi}{\partial t} = K\frac{\partial^2\psi}{\partial y^2} + \frac{(P_S E \cos\varphi_0 + M)}{2}\psi(\Theta_0^2 - \psi^2) \tag{8}$$

where $K\partial^2\psi/\partial y^2$ and $\gamma_\psi \partial\psi/\partial t$ are the elastic and viscous moments defining the director reorientation by the angle ψ; K and γ_ψ are elasticity and viscosity coefficients; t is the time; and the electric field switches on at $t = 0$.

If we introduce variables

$$\alpha = \frac{d^2(P_S E \cos\varphi_0 + M)}{2K}, y' = \frac{y}{d}, t' = \frac{t}{t_0}, \text{ and } t_0 = \frac{\gamma_\psi d^2}{K}, \tag{9}$$

then Equation (8) will be transformed to

$$\frac{\partial\psi}{\partial t'} + \frac{\partial^2\psi}{\partial y'^2} + \alpha\Theta_0{}^2\psi - \alpha\psi^3 = 0. \tag{10}$$

Since ψ in a general case can depend not only on the coordinates but also on the time, we have $\psi = \psi(y', t') = \psi_0 \exp(-i\xi t')$, where $\xi > 0$ is a constant, and ψ_0 is the amplitude, which is a slow function of time [24]. Thus, Equation (8) is converted to the following:

$$\frac{\partial\psi_0}{\partial t'} + \frac{\partial^2\psi_0}{\partial y'^2} + \psi_0\left(\alpha\Theta_0{}^2 + i\xi\right) - \alpha\psi_0{}^3 = 0. \tag{11}$$

After substitution of $\psi_0 = \Phi\exp(-i\eta)$, where $\Phi = \Phi(y')$ and $\eta = \eta(t')$, Equation (11) is transformed to a system of two equations:

$$\begin{cases} \partial^2\Phi/\partial y'^2 + \Phi\alpha\Theta_0{}^2 - \alpha\,\Phi^3 = 0, \\ \Phi(\partial\eta/\partial t') - \Phi\xi = 0. \end{cases} \tag{12}$$

Using the hyperbolic function *sn*, one can describe Φ as follows:

$$\Phi = \frac{\sqrt{2}\Theta_0 k}{\sqrt{1+k^2}} sn\left[\frac{\Theta_0\sqrt{\alpha}(y'+C_1)}{\sqrt{1+k^2}}, k\right], \text{ and } \eta = \xi t' + C_2. \tag{13}$$

Now the function $\psi = \psi_0\exp(-i\xi t')$ using *sn* can be written as follows:

$$\psi = \frac{\sqrt{2}\Theta_0 k}{\sqrt{1+k^2}} sn\left[\frac{\Theta_0\sqrt{\alpha}(y'+C_1)}{\sqrt{1+k^2}}, k\right]\exp\left(-2i\xi t' - iC_2\right). \tag{14}$$

For the extremely nonlinear situation at $k \to 1$ when the wave period tends to infinity, this relation (using the hyperbolic function *th*) results in the spatially localized waves of a stationary profile—solitons.

$$\psi = \Theta_0 th\left(\frac{\Theta_0\sqrt{\alpha}(y'+S_1)}{2}\right)\exp(-2i\xi t' - iC_2) \tag{15}$$

From the boundary and initial conditions, we take $C_1 = C_2 = 0$ and $\eta = $ const. Then,

$$\psi = \Theta_0 th\left(\frac{\Theta_0\sqrt{\alpha}y'}{2}\right)\exp(-2i\xi t'). \tag{16}$$

This equation defines the width of the soliton localization region and its amplitude. However, it does not describe the soliton movement along the coordinate y, namely, its velocity V. This is defined through the transformation

$$\psi_0 = \Phi(y' - Vt')\exp[i(V/2)(y' - (V/2)t')]. \tag{17}$$

Then, the function ψ is

$$\psi = \sqrt{\frac{2}{\alpha}} \frac{\sqrt{\alpha\Theta^2 - V^2/4}k}{\sqrt{1+k^2}} sn\left[\frac{\sqrt{\alpha\Theta^2 - V^2/4}y'}{\sqrt{1+k^2}}, k\right] \exp\left(i\frac{V}{2}(y' - Vt') - i\omega t'\right), \qquad (18)$$

where $\omega = \xi - V^2/4$ is the frequency in the system of reference moving with a soliton, and ξ is the frequency at the fixed (laboratory) system of reference.

From Equation (18) at $k\rightarrow 1$, we finally obtain a spatially localized solution—the wave of the stationary profile moving with velocity V along the coordinate y:

$$\psi = \sqrt{\frac{\alpha\Theta^2 - V^2/4}{\alpha}} th\left[\sqrt{\frac{\alpha\Theta^2 - V^2/4}{2}}(y' - Vt')\right] \exp\left(i\frac{V}{2}(y' - Vt') - i\omega t'\right). \qquad (19)$$

So, the spatially localized solution of Equation (8) is a two-parameter soliton: the first parameter is the velocity V of its center motion, and the second one is its eigenfrequency ω in the system of reference moving with the soliton. Such a solution is presented in graphical form in Figure 7.

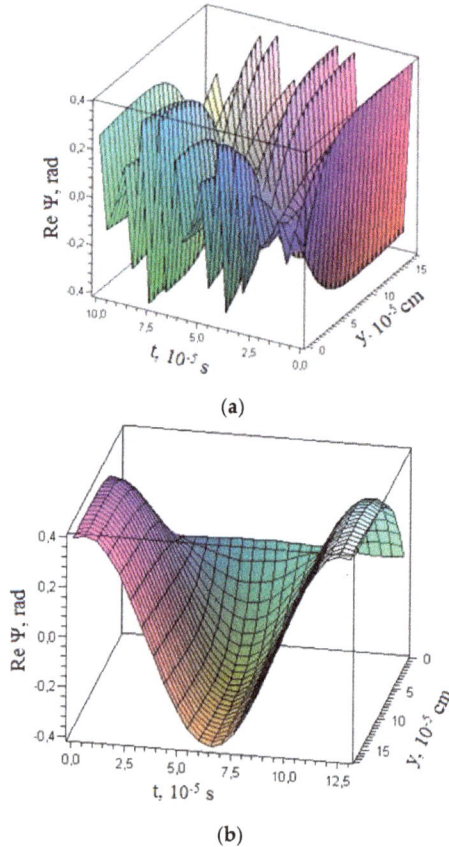

(a)

(b)

Figure 7. Graphical presentation of the solution of Equation (18), which describes the change of the angle ψ in the electric field E (**a**) and motion of the orientation bend in the FLC layer (**b**). $E = 3$ V/μm, $P_S = 50$ nC/cm², $M = 4 \times 10^3$ erg/cm³, $K = 5 \times 10^{-12}$ N, $\gamma_\psi = 0.2$ P, $\Theta_0 = 23°$, and $\varphi_0 = 30°$.

As seen from Figure 7, the transition to the Maxwellian mechanism of the energy dissipation when the dissipative coefficient is the shear viscosity γ_ψ results in the appearance of a soliton, which is a wave packet with a periodic wave localized therein. The maximum speed of soliton motion found from Equations (9) and (19) is

$$V = 2\Theta_0 \sqrt{\alpha K / (\gamma_\psi d)} = (\Theta_0 / \gamma_\psi) \sqrt{2K(M + P_S E \cos \varphi_0)}, \qquad (20)$$

and the time of director reorientation due to the movement of the orientation bend (Figure 7b) is

$$\tau_C = \frac{\gamma_\psi d^2}{K\xi} = \frac{2\gamma_\psi}{\Theta_0^2 (P_S E \cos \varphi_0 + M)}. \qquad (21)$$

If $\varphi_0 = 30°$, $P_S = 50$ nC/cm^2, $M = 4 \times 10^3$ erg/cm^3, $K = 5 \times 10^{-12}$ N, $E = 3$ V/μm, $\Theta_0 = 23°$, and $\gamma_\psi = 0.2$ P, then the speed of the soliton center motion is $V = 0.65$ cm/s, and the director reorientation time τ_C is of about ≈ 150 μs.

6. Light Modulation in an Electro-Optic Cell with Helix-Free Ferroelectric Liquid Crystal (Experimental Results)

6.1. Modulation of Light Transmission

The modulation of light transmission was observed when an electro-optic cell of 1.7 μm thickness (achromatic for visible light) was placed between crossed polarizers and bipolar voltage of the rectangular shape (meander) was applied. The FLC director reorientation due to motion of solitons made it possible to reduce the optical response time in weak fields to 25 μs and to reach the light modulation frequency of 7 kHz at the control voltage of ± 1.5 V [3] (Figure 8).

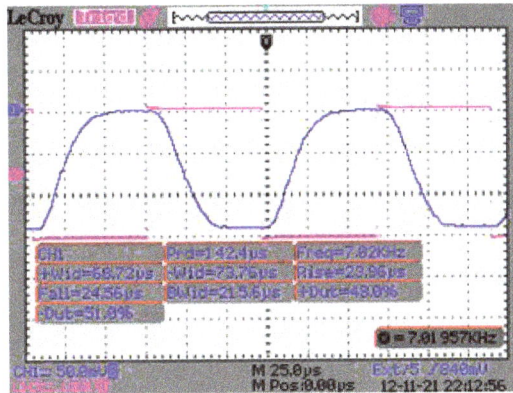

Figure 8. Oscillogram of the bipolar control voltage (zero level—digit 3) and optical response (zero level—digit 1) for an electro-optical cell with composition HF-32C at the voltage amplitude of ± 1.5 V and frequency of 7.0 kHz. The upper level of the optical response is the closed state; the lower one is the light transmission state.

The transition to the soliton mode can occur not only with increasing frequency of the electric field change, but also with increasing field strength at fixed frequency. This transition is accompanied by a sharp decrease in the optical response time when a certain threshold value of the field strength is reached (Figure 9). With increasing frequency, the threshold value of the field for the transition to the soliton mode decreases, and the time of the optical response $\tau_{0.1-0.9}$ decreases (Figure 9, curve 2) at a lower frequency.

After the transition to the soliton mode, a section in the dependence $\tau_{0.1-0.9}$ (*E*) appears where the optical response time is independent of the electric field strength (Figure 9). This means that the FLC dissipative coefficient is the shear viscosity.

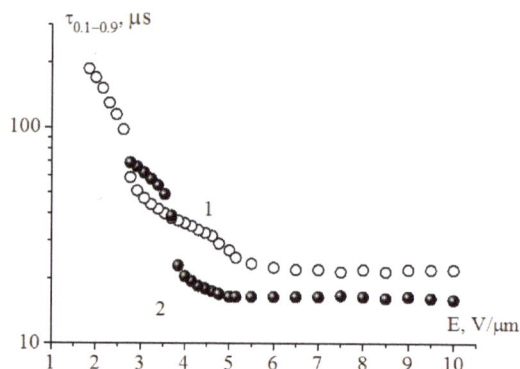

Figure 9. Field dependences of the optical response time for a cell with the HF32F composition. The thickness of the electro-optical cell is 1.7 μm. The voltage frequencies (meander) are 200 Hz (curve 1) and 3 kHz (curve 2).

Constant change in the director position along the FLC smectic layers provides the substantially hysteresis-free dependence of the light transmission of the electro-optical cell on the control voltage amplitude when both increasing and decreasing in a wide frequency range. Experiments have shown that the hysteresis-free modulation characteristic I (V) for both positive and negative voltage occurs if the following two conditions are satisfied: First, the frequency control voltage corresponds to the frequency interval of existence of the soliton mode (for the HF-32C composition, it extends from 100 Hz to 7 kHz). Second, this frequency does not correspond to the static portions (Figure 10) of the frequency dependence of the FLC birefringence Δn (f). This means that the frequency of the hysteresis-free modulation does not exceed 6 kHz (Figure 11). This maximum value was realized experimentally [25].

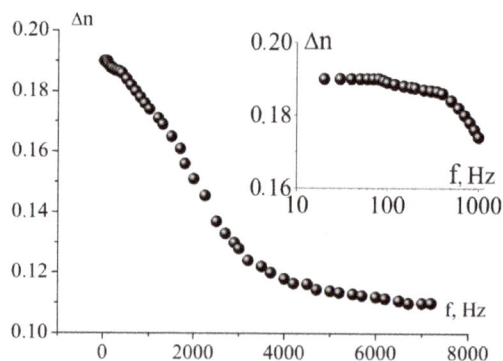

Figure 10. Frequency dependence of the birefringence of the HF-32C composition. Inset: a low-frequency part of this dependence. The thickness of the electro-optical cell is 1.7 μm. The amplitude of the bipolar control voltage (meander) is ±1.5 V.

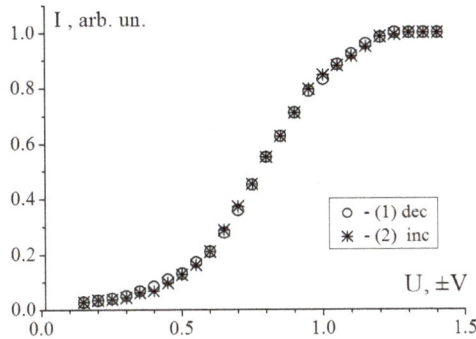

Figure 11. Dependence of the light transmission on the electrical voltage for the frequency 6 kHz at decreasing (1) and increasing (2) voltage. The thickness of the electro-optical cell with the HF-32C composition is 1.7 μm.

Note that the halftone modulation characteristic in such a uniquely wide frequency range is realized only on the basis of the physical properties of new FLC materials. In addition, unlike the bistable (two-level) characteristic of the known FLCs, possible applications of novel helix-free FLC do not require additional electronic modulation of a signal, which reduces the frame rate of color image formation.

It is important that the predominance of shear viscosity in a soliton mode results in weakening the temperature dependence of the electro-optical response time in a wide temperature interval (Figure 12). The higher the frequency of light modulation (and the frequency of the control voltage), the wider the temperature interval in which the time $\tau_{0.1-0.9}$ is almost constant. It shifts to both high and low temperatures, and for the HF-32C composition, the time $\tau_{0.1-0.9}$ at the maximum possible light modulation frequency of 7 kHz is from 10 to 52 °C.

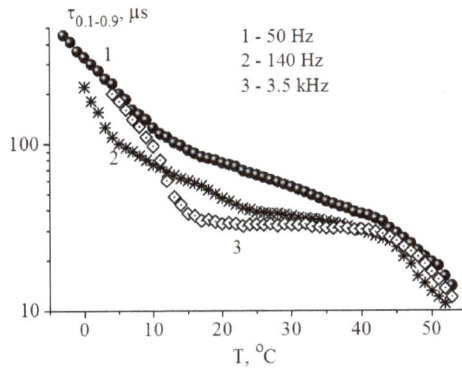

Figure 12. Temperature dependences of the electro-optical response time for the 1.7 μm thick FLC cell with the composition HF-32B. The amplitude of the control voltage (meander) is ±1.5 V. The control voltage frequencies are 50 Hz (curve 1), 140 Hz (curve 2), and 3.5 kHz (curve 3).

The experimental results show that new helix-free FLCs are very promising materials for the next generation of LC displays, especially for fast displays using FLC on Silicon and 3D technologies, as well as the progressive Field Sequential Colors (FSC) technique, which allows us to form brighter images (because of the absence of RGB filters) by a display with a 3-times-smaller number of pixels [26,27].

6.2. Modulation of Light Scattering

The deformation of a single-domain structure of the FLC caused by a pulse of the electric field can be accompanied under certain conditions by a short "flash" of light scattering. Such scattering was observed for the first time in 1984 and was called "transient" [4], but it was considered parasitic and was not studied in detail. Intense transient scattering in the helix-free FLC was first studied in [30], where it was also proposed to be used in high-speed modulators of a three-dimensional display with a volumetric screen (volumetric display).

Transient light scattering in new helix-free FLCs, including bistable scattering, was studied in detail in [5,16]. This scattering occurs on the boundaries of spontaneously ordered regions which are formed in the process of the appearance of waves of a stationary profile, i.e., solitons. Scattering occurs after changing the electric field sign and disappears when the motion of solitons reorients the director in all smectic layers—that is, a new homogeneous structure of the FLC layer is no longer formed. A change in the electric field direction induces transient domain formation again, and the process is repeated.

The frequency dependence of the optical response time for light scattering is similar to that described in Section 4. The transition to the Maxwellian mechanism of energy dissipation is accompanied by a strong frequency dependence of the response time $\tau_{0.1-0.9}$ similar to that shown in Figure 4. Some increase in the time $\tau_{0.1-0.9}$ is due to the simultaneous presence of two dissipative coefficients γ_φ and γ_ψ. After the transition to the soliton mode, the time $\tau_{0.1-0.9}$ is determined by the velocity of soliton wave motion, so the frequency dependence of the response time is practically absent.

During operation in a light-scattering mode, the polarizers are not required, and this increases the light transmission of the electro-optical cell by up to 80%. Basically, it is limited by the transparency of conductive coatings on glass substrates. If the pulse duration is less than the minimum time required for a complete disappearance of the transition domains, the light transmission of the cell decreases.

The maximum efficiency of light scattering corresponds to the regular structure of the scattering centers in the form of circular domains that are fairly uniformly distributed throughout the volume of the FLC layer. A decrease in the cell thickness shifts the maximum corresponding to a regular scattering structure toward shorter pulse durations, but the contrast ratio also decreases.

For a certain experimentally chosen relationship between the amplitude and duration of the alternating impulses of the control voltage, and also between the elastic deformation energy of smectic layers and the FLC spontaneous polarization, the process of light scattering on the dynamic domain structure at the transition to the Maxwellian energy dissipation becomes bistable with a maximum light transmission above 80% and a contrast ratio of about 200:1 (Figure 13). Both optical states (with or without scattering) could be turned off for a few tens of microseconds and be memorized for a few tens of seconds, or until a pulse of opposite polarity was applied. The maximum light-scattering modulation frequency was about 5 kHz [16].

A change in the duty cycle between the control voltage pulses (the duration is maintained) leads to a change in the ratio between the lifetimes of both optical states. When an alternating pulse duration is reversed (the duty cycle is maintained), and the duration of a pulse switching on the scattering becomes equal to the duration of a pulse switching off the scattering (and vice versa), the ratio of the lifetimes of states with the maximum light transmission and with the maximum light scattering is reversed.

There is a limit on the minimum duration of voltage pulses which turn on and turn off the scattering process. It should not exceed the characteristic reorientation time of the director caused by the motion of the orientation inflection [16]. From Equation (21), this is about 150 μs for the parameters indicated there.

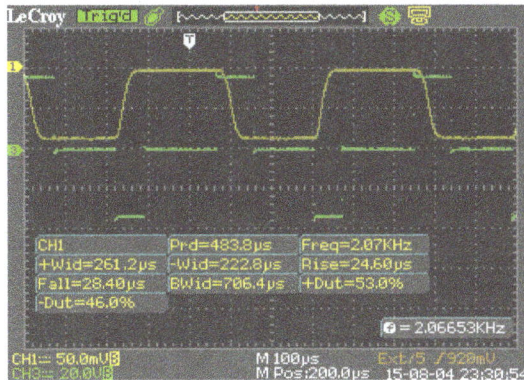

Figure 13. Oscillograms of the control voltage (zero level—digit 3) and optical response (zero level—digit 1) for the bistable switching mode. The upper level of the optical response is the scattering state, the lower level is the nonscattering state. The thickness of the FLC cell with the HF-32 composition is 13 μm. Control voltage—bipolar pulses of ±35 V. The frequency of light modulation is 2 kHz.

There is an optimal relationship between the period of deformation of FLC smectic layers and the electro-optical cell thickness when, at certain electric field strength, the velocity of the soliton wave motion is maximal (the optical response time $\tau_{0.1-0.9}$ is minimal) and the light modulation frequency is also maximal.

Depending on the time of the electric field action (duration of voltage pulses) and the cell thickness, there may occur several maxima of light scattering that may be treated as the scattering efficiency C (or contrast ratio) (Figure 14). The emergence of the second and third maxima of light scattering efficiency occurs with increasing the cell thickness up to 16–20 μm [16].

The maximum efficiency of light scattering and the maximum light transmittance without scattering are achieved at different durations of control voltage pulses. An increase in the pulse duration results in increasing the domain wall length and leads to irregular scattering structures. As a result, the density of scattering centers reduces; this is a reason for the light scattering efficiency decrease (Figure 14).

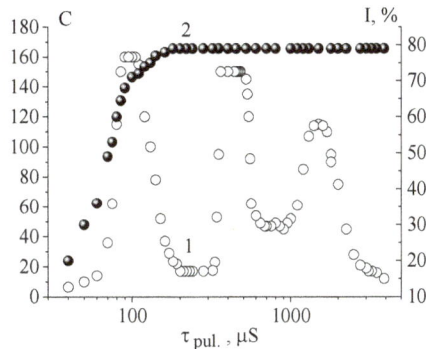

Figure 14. Dependences of light scattering C (curve 1) and light transmission I without scattering (curve 2) on the duration of bipolar voltage pulses with a fixed amplitude (±50 V). The thickness of an FLC cell with the HF-32 composition is 18 μm.

After switching off the electric field, the transmitted optical radiation does not change in spectral composition, and there is no residual light scattering in the FLC layer (due to the absence of a helix). Since the magnitude of the spontaneous polarization does not exceed 50 nC/cm^2, ferroelectric domains do not arise. Therefore, in the absence of an electric field, scattering and diffraction centers are absent. The saturation voltage is rather low and, consequently, the control voltage of an electro-optical cell is also quite small (less than 50 V).

The main possible applications of light-scattering FLC compositions are the following: polarizer-free visible and infrared optical shutters, energetically effective screens of electronic books, 3D visualizers (volumetric screens) of volumetric displays, etc. [16,27].

6.3. Spatially Inhomogeneous Modulation of Light Phase Delay

When the duration of voltage pulses supplied simultaneously to the electro-optical cell corresponds to different maxima of light scattering efficiency, the transitions between light-scattering modes (which correspond to light scattering maxima) result in the most chaotic changes in the position of a scattering indicatrix. As a result of a short-term switching on of light scattering (less than for 50 μs), structures with an almost random distribution of refractive index gradients are formed in the entire volume of the FLC. They cause a spatially inhomogeneous phase modulation of the light beam (over its cross section) passing through the electro-optical cell.

Spatially inhomogeneous modulation with a phase delay of the order of and more than π allows one to destroy the phase relations in a laser beam passing through the electro-optical cell and, further, to suppress the speckle noise in images formed due to the ability of laser rays to interfere.

This approach was used to develop the first electro-optical despeckler, the device that reduces the contrast of speckles and, due to this, suppresses the speckle noise in laser images. In [28,29], helix FLCs were used for this aim, and an alternating electric field was applied across the cell simultaneously at low and high frequencies, which caused spatial deformations of the helicoid in an FLC layer. Unfortunately, such a despeckler had serious drawbacks. First, the deformed helix structure of molecules changed the spectral composition of the laser radiation. Besides this, after the electric field was switched off, the residual scattering caused by the helices remained. Also, the light modulation frequency at the electric field strength of ~2 V/μm was limited by the value of 500 Hz, which hampers possible applications of this despeckler.

These drawbacks were eliminated after using new helix-free FLC in the experimental samples of a despeckler. Figure 15 illustrates the operation of one of the best samples. Signals of the control voltage and the modulated optical response are shown in Figure 15a. To create the different nonrepeating distributions of refractive index gradients in the volume of the FLC layer, two-frequency voltage pulses (meander) were supplied to the electro-optical cell, and pulses of low frequency (2 kHz) were modulated by pulses of high frequency (10 kHz).

In Figure 15b,c, the speckle intensity distributions in the cross section of the laser beam behind the FLC electro-optical cell are presented for when the control voltage on the cell electrodes is not applied and is applied, respectively. The optical data were recorded using a CCD (charge-coupled device) camera and were processed using a special software product. The reduction in the contrast of the speckle pattern was calculated from the data in Figure 15b,c as the ratio R between the contrast C_0 of a speckle pattern measured without the despeckler sample and the contrast C_1 of a speckle pattern measured with the despeckler sample: R = 10 $\log_{10}$ (C_0/C_1) = 10 $\log_{10}$ (0.82/0.07) = 10.2 dB [30,31].

The experiments confirmed the simplicity and high efficiency of the electro-optical despeckler prototype based on an electro-optical cell with the helix-free FLC. Such a device can be widely used in holography and in laser projection displays.

(a)

(b)

(c)

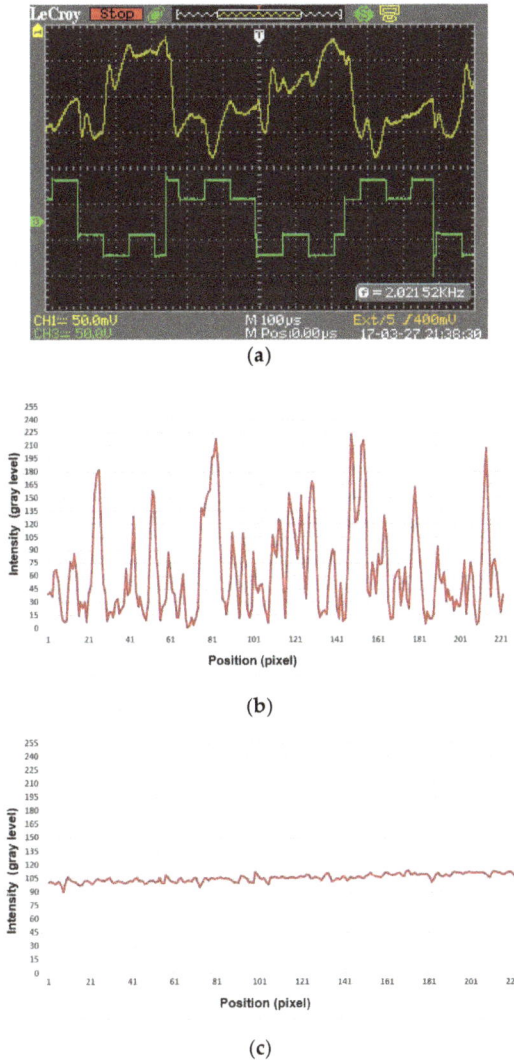

Figure 15. Illustration of the operation of an experimental sample of a despeckler using the novel helix-free FLC: (**a**) Oscillograms of the control voltage (zero level—digit 3) and phase delay modulation (zero level—digit 1); value of a low-frequency signal (meander) 2 kHz and amplitude ± 50 V; frequency of the modulating signal is 7 kHz and amplitude ± 15 V; electric field strength is 3.25 V/µm; (**b**) Radiation intensity distribution in the cross section of a laser beam transmitted through the electro-optical despeckler sample when the control voltage is zero; contrast of the speckle pattern $C_0 = 0.82$; (**c**) Radiation intensity distribution in the cross section of a laser beam in the presence of the control voltage; contrast of the speckle pattern $C_1 = 0.07$. The thickness of the FLC cell is 20 µm. The wavelength of the laser radiation is 0.65 µm.

7. Conclusions

Based on our research results, we considered in detail the physical properties, light modulation characteristics, and possible applications of novel helix-free FLC developed and fabricated at the

Lebedev Physical Institute, Moscow. They are distinguished by the spatially periodic deformation of smectic layers, a small value of spontaneous polarization, rather high viscosity, and the soliton mechanism of FLC director reorientation upon transition to the Maxwellian mechanism of energy dissipation.

For FLC with a viscosity in the range from 0.15 to 1.0 P, the frequency and field dependences of the electro-optical response time were studied under modulation of light transmission, light scattering, and light phase delay.

When modulating the light transmission in an electric field of the order of 1 V/μm (at the control voltage of $\pm$1.5 V), the smallest response time (25 μs) and the largest modulation frequency interval (7 kHz, including 6 kHz for hysteretic-free modulation with continuous grayscale) were realized experimentally for the first time.

A theoretical model was proposed that described satisfactorily the deformation of smectic layers in the absence of the external electric field and the soliton mechanism of FLC director reorientation under an alternating electric field.

In certain compositions of the new FLC, intensive light scattering on transient domains was realized, including the discovery for the first time of bistable scattering with a memory time that can exceed the optical switching time (tens of microseconds) by 6 orders of magnitude (up to tens of seconds).

Initiated by the short-time switching on of light scattering, the spatially inhomogeneous phase light modulation due to random small-scale gradients of the FLC refractive index was studied. This modulation is capable of destroying the phase relations in a laser beam and of suppressing the speckle noise in images formed by the beam. Based on this, experimental samples of simple and effective electro-optical despecklers were studied for the first time.

Possible applications of the novel helix-free FLC are the fastest display devices (including displays using FLCoS, 3D, and FSC technologies), spatial light modulators, light-scattering polarizer-free modulators (including infrared devices and optical-state-memorizing ones), energetically effective screens of electronics books, 3D visualizers (volumetric screens) of volumetric displays, coding–decoding devices, etc.

8. Patents

Two U.S. patents were issued based on results of the research described above, namely

- Patent No. US 9,532,037 B2: Fast-Acting Low Voltage Liquid Crystal Stereo Glasses. Inventors: Kompanets I.N., Andreev A.L., Ezhov V.A., Sobolev A.G. Date of Patent: 27 December 2016 (PCT Pub. Date: 22 May 2014).
- Patent No. US 9,709,877 B2: Video Projector Employing Ferroelectric Liquid Crystal Display. Inventors: Kompanets I.N., Andreev A.L. Date of Patent: 18 July 2017 (PCT Pub. Date: 5 December 2013).

Author Contributions: Conceptualization, A.A. and I.K.; methodology, A.A.; software, N.Z.; theoretical model, A.A. and T.A.; investigation, A.A. and N.Z.; resources, I.K.; data curation, N.Z.; writing and editing, I.K.; supervision and project administration, I.K.; funding acquisition, I.K.

Funding: This research was funded by the Ministry of Education and Science of the Russian Federation; the unique identifier of the project is RFMEFI60417X0191.

Acknowledgments: The authors thank A. V. Novozhenov for the assistance with this work, Yu. P. Bobylev and V. M. Shoshin for preparing the FLC cells as well as I. Revokatova for technical support.

Conflicts of Interest: The authors declare no conflict of interest.

References

1. Lagerwall, S.T. *Ferroelectric and Antiferroelectric Liquid Crystals*; WILEY-VCH Verlag GmbH: Weinheim, Germany, 1999; pp. 241–257. ISBN 3527298312.

2. Clark, N.A.; Lagerwall, S.T. Sub-microsecond switching in ferroelectric liquid crystals. *J. Appl. Phys.* **1980**, *36*, 899–903.

3. Chigrinov, V.G. *Liquid Crystal Devices: Physics and Applications*; Artech House: London, UK, 1999; 427p, ISBN 13 9780890068984.

4. Katsumi, Y.; Ozaki, M. New electrooptic effect of microsecond response utilizing transient light scattering in ferroelectric liquid crystal. *J. Appl. Phys. Jpn.* **1984**, *23*, L385–L387.

5. Andreev, A.L.; Bobylev, Y.P.; Kompanets, I.N.; Pozhidaev, E.P.; Fedosenkova, T.B.; Shoshin, V.M.; Shumkina, Y.P. Electrically controlled light scattering in ferroelectric liquid crystals. *J. Opt. Technol.* **2005**, *72*, 701–707. [CrossRef]

6. Beresnev, L.A.; Baykalov, V.A.; Blinov, L.M.; Pozhidaev, E.P.; Purvanetskas, G.V. First non-helix ferroelectric liquid crystal. *J. Pisma v ZhETF* **1981**, *33*, 553–556.

7. Ostrovsky, B.I.; Chigrinov, V.G. Linear electro-optic effect in chiral smectic C * liquid crystals. *Crystallography* **1980**, *25*, 322–331.

8. Handschy, M.A.; Clark, N.A.; Lagerwall, S.T. Field-Induced First-Order Orientation Transitions in Ferroelectric Liquid Crystals. *Phys. Rev. Lett.* **1983**, *51*, 471–474. [CrossRef]

9. Andreev, A.L.; Kompanets, I.N.; Andreeva, T.B.; Shumkina, Y.P. Dynamics of domain wall motion in ferroelectric liquid crystals in an electric field. *J. Phys. Solid State* **2009**, *51*, 2415–2420. [CrossRef]

10. Landau, L.D.; Lifshits, E.M. *Theory of Elasticity*; Publisher: Nauka, Moscow, 1987; pp. 188–189. (In Russian)

11. Andreev, A.; Andreeva, T.; Kompanets, I. Fast Low Voltage FLC Materials for Active Matrix Displays. In Proceedings of the 29th IDRC (Eurodisplay-09), Rome, Italy, 14–17 September 2009; Dalaad Edizioni: Rome, Italy, 2009; pp. 366–369.

12. Andreev, A.; Andreeva, T.; Kompanets, I. Low Voltage FLC for Fast Active Matrix Displays. In Proceedings of the SID'10 Symposium Digest, Seattle, WA, USA, 23–28 May 2010; Volume 41, pp. 1716–1719.

13. Andreev, A.L.; Andreeva, T.B.; Kompanets, I.N. Electro-Optical Response of Compensated Helix Ferroelectric: Continuous Gray Scale, Fastest Response and Lowest Control Voltage demonstrated to date. In Proceedings of the SID'12 Symposium Digest, Boston, MA, USA, 3–8 June 2012; Volume 43, pp. 452–455.

14. Andreev, A.L.; Andreeva, T.B.; Kompanets, I.N.; Zalyapin, N.V. Increasing the light modulation frequency due to the increase of FLC viscosity. In Proceedings of the SID'13 Symposium Digest, Vancouver, BC, Canada, 19–25 May 2013; Volume 44, pp. 1303–1306.

15. Andreev, A.L.; Andreeva, T.B.; Kompanets, I.N.; Zalyapin, N.V. Optical response time of helix-free СЖК: Continuous gray scale, fastest response, and lowest control voltage. *J. SID* **2014**, *22*, 115–121. [CrossRef]

16. Andreev, A.; Andreeva, T.; Kompanets, I.; Zalyapin, N.; Xu, H.; Pivnenko, M.; Chu, D. Fast bistable intensive light scattering in helix-free ferroelectric liquid crystals. *Appl. Opt.* **2016**, *55*, 3483–3492. [CrossRef] [PubMed]

17. Abdulhalim, I.; Moddel, G.; Clark, N.A. Director-polarization reorientation via solitary waves in ferroelectric liquid crystals. *Appl. Phys. Lett.* **1992**, *60*, 551–553. [CrossRef]

18. Fedosenkova, T.; Andreev, A.; Pozhidaev, E.; Kompanets, I. Birefringence controlled by external electric field in helix-free ferroelectric liquid crystals. *Bull. Lebedev Phys. Inst.* **2002**, *3*, 36–42.

19. Maclennan, J.E.; Clark, N.A.; Handschy, M.A. Solitary waves in ferroelectric liquid crystals. In *Solitons in Liquid Crystals*; Lam, L., Prost, J., Eds.; Springer: New York, NY, USA, 1991; pp. 151–190. ISBN 0941-5114 or 9780412754500.

20. Akhmediev, N.; Ankiewich, A. *Solitons, Nonlinear Pulses and Beams*; Chapman & Hall: London, UK; New York, NY, USA; Tokyo, Japan; Melbourne, Australia; Madras, India, 1997; 335p, ISBN 0412754509.

21. Song, J.K.; Sufin, M.J.; Vij, J.K. Solitary wave propagations in surface stabilized ferroelectric liquid crystal cells. *Appl. Phys. Lett.* **2008**, *92*, 083510. [CrossRef]

22. De Gennes, P.G. *Physics of Liquid Crystals*; Clarendon Press: Oxford, UK, 1974; 616p, ISBN 0198520247.

23. Pavel, J.; Glogarova, M. A new type of layer structure defects in chiral smectics. *Liquid Cryst.* **1991**, *9*, 87–93. [CrossRef]

24. Kosevich, A.M.; Kovalev, A.S. *Introduction to Nonlinear Physical Mechanics*; Naukova Dumka: Kiev, Ukraine, 1988.

25. Andreev, A.L.; Andreeva, T.B.; Kompanets, I.N.; Zalyapin, N.V. Hysteresis-Free Modulation Characteristic and Electro-Optical Response in Helix-Free Ferroelectric Liquid Crystals. In Proceedings of the International Conference on Display Technology (ICDT'2018), Guangzhou, China, 9–12 April 2018; pp. 361–364.

26. O'Callaghan, M.J.; Handschy, M.A. Ferroelectric liquid crystal SLMs: From prototypes to products. *Proc. SPIE* **4457**, *4457*, 31–42.
27. Andreev, A.L.; Andreeva, T.B.; Kompanets, I.N.; Zalyapin, N.V.; Starikov, R.S. Novel FLC materials open new possibilities for FLCOS microdisplays and video projectors. *Phys. Procedia* **2015**, *73*, 87–94. [CrossRef]
28. Andreev, A.L.; Andreeva, T.B.; Kompanets, I.N.; Minchenko, M.V.; Pozhidaev, E.P. Suppressing the speckle-noise using a liquid crystal cell. *Quantum Electron.* **2008**, *38*, 1166–1170. [CrossRef]
29. Andreev, A.L.; Andreeva, T.B.; Kompanets, I.N.; Minchenko, M.V.; Pozhidaev, E.P. Spekle-noise suppression due to a single ferroelectric liquid crystal cell. *J. SID* **2009**, *17*, 801–807. [CrossRef]
30. Zalyapin, N.V.; Andreev, A.L.; Andreeva, T.B.; Kompanets, I.N. An electro-optical despeckler based on the helix-free ferroelectric liquid crystal. *Quantum Electron.* **2017**, *47*, 1064–1068. [CrossRef]
31. Andreev, A.L.; Andreeva, T.B.; Kompanets, I.N.; Zalyapin, N.V. Space-inhomogeneous phase modulation of laser radiation in an electro-optical ferroelectric liquid crystal cell for suppressing speckle noise. *Appl. Opt.* **2018**, *57*, 1331–1337. [CrossRef] [PubMed]

applied
sciences

MDPI

Article

Phase-Only Optically Addressable Spatial-Light Modulator and On-Line Phase-Modulation Detection System

Lili Pei, Dajie Huang *, Wei Fan *, He Cheng and Xuechun Li

National Laboratory on High Power Laser and Physics, Shanghai Institute of Optics and Fine Mechanics, Chinese Academy of Sciences, Shanghai 201800, China; lilypei@siom.ac.cn (L.P.); chenghe@siom.ac.cn (H.C.); lixuechun@siom.ac.cn (X.L.)
* Correspondence: hdajie@siom.ac.cn (D.H.); fanweil@siom.ac.cn (W.F.)

Received: 3 September 2018; Accepted: 27 September 2018; Published: 3 October 2018

Abstract: The influence of driving conditions on the phase-modulation ability of an optically addressable spatial-light modulator (OASLM) is investigated using an equivalent circuit method and a system for measuring wave-front modulation that uses a phase-unwrapping data-processing method, and is constructed with a charge-coupled device and wave-front sensor. 1λ peak-to-valley phase change for a 1053 nm laser beam is acquired with the home-made OASLM at the optimal driving voltage of 14 V at 200 Hz. The detection system for wave-front modulation has a spatial resolution of 200 µm for binary images and a minimum distinguishable contrast of 1 mm. On-line phase modulation with feedback control can be acquired with the OASLM and the corresponding measuring system.

Keywords: spatial light modulator; phase change; spatial resolution

1. Introduction

Liquid crystals (LCs) [1], as tunable elements, are increasingly being used for non-display applications, including intelligent windows, tunable phase-retarders, terahertz bandgap fibers, and spatial-light modulators [2–5]. Owing to the advantages of low cost and high compactness, phase-only spatial light-modulators (SLMs) have additional potential applications in areas such as adaptive optics and wave-front control [6–8]. Two of the most common are liquid crystal on silicon (LCOS) and optically addressable types. The LCOS type is reflective and is an electrically addressed type, which means it can easily cause spectral distortion because of Fabry–Perot interference and the black-matrix effect generated by the two-dimensional (2D) periodic opaque electrodes, which can make the beam quality worse. Compared to an electrically addressable spatial-light modulator, the optically addressable spatial-light modulator (OASLM) does not require complex addressable electrode circuits. The filling factor of this type reaches 100%, which means that the generated phase distribution is continuous, not multistep. Its high resolution and simple manufacturing process renders OASLM an ideal alternative for wave-front modulation of laser beams [9].

An OASLM with parallel-aligned liquid-crystal light valves (LCLVs) works in phase-only mode and the influence of driving conditions on the phase-modulation capability of the OASLM is decisive for its performance. Traditional measurement of the phase retardation of a laser is performed using a Mach–Zehnder interferometer [10], Twyman–Green interferometer [11], and wave-front sensor (WFS) [12]. The first two devices are very sensitive to mechanical vibration and air turbulence, and the WFS is limited by its inherently restrictive measurement accuracy range. López-Téllez et al. proposed the unwrapping method to characterize the retardance function of liquid-crystal variable retarders (LCVRs) [13]. This method can quickly and accurately obtain the original, continuous function of the voltage-retardance relationship by removing discontinuities and is applicable to measure the retardance variation of the SLM.

In this study, the impact of driving conditions, including the frequency of the applied voltage and the write-light irradiance on the modulation ability of phase-only OASLM, is investigated. 1λ phase-control ability can be acquired using the home-made OASLM under an applied voltage of 14 V at 200 Hz. In order to monitor the phase-control ability of the OASLM, a detection system including a charge-coupled device (CCD) and WFS was constructed and the aforementioned unwrapping procedure was applied to deduce the phase modulation from the transmittance. The modulation result of a binary image with variable frequencies, and the modulation transfer function (MTF), indicate that the proposed system has a spatial resolution of 200 μm for modulation measurement and a minimum distinguishable contrast of 1 mm. This detection system can correctly monitor the modulation results of the 2D chessboard grayscale image.

2. The Influence of the Driving Conditions on the Performance of a Phase-Only OASLM

The phase-only OASLM is schematically depicted in Figure 1a. Considering the photoconductive characteristic and spectral absorption coefficient of $Bi_{12}SiO_{20}$ (BSO) crystal [14], we use the light-emitting diode (LED) with a wavelength of 470 nm as the light source of the write light. An amplitude-modulating LCOS SLM (Rui Like Co., WXGA Active Matrix LCD, the pixel pitch is 20 μm) has been used to control the collimated beam from the LED source. The bitmapped image is then projected onto the BSO layer through an imaging system. We can control the wave-front of the transmissive read light (with a wavelength of 1053 nm the photoconductive effect of the BSO crystal at 1053 nm is almost negligible) by setting an appropriate bitmap in the LCOS modulator.

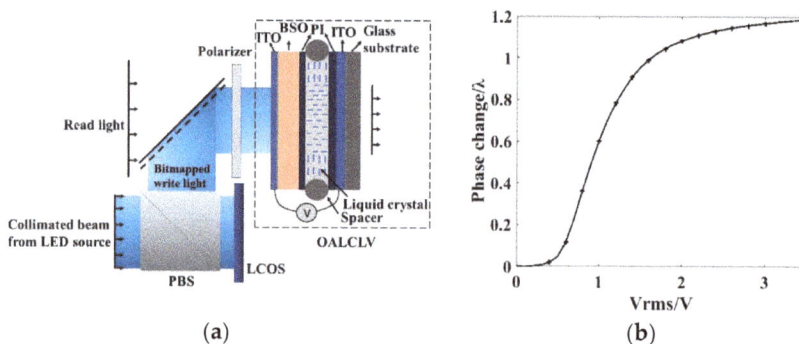

Figure 1. (a) Working principle of the optically addressable spatial-light modulator (OASLM) and structure of the optically addressable liquid-crystal light valve (OALCLV) consisting of the glass substrate, 6-μm-thick liquid crystal (LC), and BSO photoconductor, PI (polyimide, the alignment layer); a reflective liquid crystal on silicon (LCOS) type is used to control the write light, PBS refers to the polarized beam splitter. (b) Phase change of a 6-μm-thick LC cell composed of two glass substrates. The LC has $n_e = 1.820$ and $n_o = 1.515$.

When the BSO layer (1 mm thickness, 20×20 mm² effective area) is illuminated with the write light, and an electric field is applied across the optically addressable liquid-crystal light valve (OALCLV), the resistance of the BSO decreases and the voltage begins to act on the LC layer, so the LC director is redirected towards the applied field [15]. As the voltage increases, the director rotates further, which changes the extraordinary refractive index of the LC and results in a reduction in the phase retardance of the OASLM [16]. In this way, the voltage on the LC layer affects the phase controllability of the OASLM.

The voltage response of the LC causes a phase change of the OALCLV, so first we studied the relationship between the phase change and the voltage of the LC cell without BSO. We oriented the test cell at 45° between crossed polarizers and applied the alternating voltage. The intensity value of the modulated 1053-nm laser was obtained using a CCD, and the average value of all pixels was taken as the effective value, which is expressed as follows [17].

$$T = \frac{I}{I_{max}} = \frac{1 - cos\delta}{2} \tag{1}$$

where T refers to the transmittance. Phase change is added to the laser beam by the OASLM according to the loaded grayscale images of the write light. The retardance δ of the liquid crystal layer is expressed as follows:

$$\delta = 2 \times sin^{-1}\left(\sqrt{T}\right) \tag{2}$$

The phase change is deduced from the transmittance according to Equation (2). Because of the periodicity, δ can only be within the range of $0 - \pi$, so in the data processing, the phase unwrapping [18] is performed to calculate the actual phase change.

We measured the voltage response curve of an LC cell with 6-μm-thick LC layer (HCCH Co., Jiangsu, China, n_e = 1.820 and n_o = 1.515 at λ = 589 nm, with the same thickness as the LC layer in OALCLV) and two glass substrates. The voltage response shown in Figure 1b will be used as a reference in the phase change-voltage relationship of the OALCLV. According to this figure, we can acquire the threshold voltage of LC molecule rotation as V_{thr} = 0.4 V and the saturated voltage as V_{sat} = 3 V, and the peak-to-valley (PV) value of the phase change is approximately 1.1λ.

The OALCLV, the main component of the OASLM, can be modelled with an equivalent circuit in Figure 2a. The resistance and capacitance of the LC layer are R_{LC} and C_{LC} respectively. The dark BSO crystal resistance is R_0. The resistance of the BSO under-uniform illumination level ϕ is R_ϕ. The value of R_1C_1 is inversely proportional to the illumination level ϕ. Similar models have been introduced in our previous work in Reference [19]. With the input voltage $V_{AC} = V_0 cos\omega t$, the voltage drop across the LC layer, V_{LC}, can be calculated using the following equation.

$$V_{LC} = V_{AC} \times \frac{\frac{1}{R_0} + \frac{1}{R_\phi} + j\omega C_{BSO} + \frac{j\omega C_1}{1 + j\omega R_1 C_1}}{\frac{1}{R_0} + \frac{1}{R_\phi} + j\omega C_{BSO} + \frac{j\omega C_1}{1 + j\omega R_1 C_1} + \frac{1}{R_{LC}} + j\omega C_{LC}} \tag{3}$$

These V_{LC} values as a function of the frequency of the V_{AC}, when the OALCLV is illuminated with different write-light irradiances, are shown in Figure 2b, and are solved using Equation (3) with an root mean square (RMS) value of V_{AC} = 3.5 V. Several physical parameters have been used in the model as follows [20]: $\rho_{LC} = 10^{10}$ Ω·cm and $\varepsilon_{LC} = 10$ are the density and relative dielectric constant of the LC, respectively, and $\sigma_0 = 10^{-14}$ Ω^{-1}cm^{-1} is the dark BSO crystal conductivity. The dotted line indicates V_{thr}. These marked points correspond to the experimental values, which are basically consistent with the simulated values and confirm the validity of Equation (3). From the figure we can see that V_{LC} decreases with increase in frequency when the OALCLV is under the same write-light irradiance. Meanwhile, V_{LC} increases with increase in write-light irradiance when the OALCLV is under the same driving voltage frequency.

The V_{LC} of the OALCLV must be greater than V_{thr} to ensure the voltage response and the maximum value of V_{LC} directly determines the PV value of the phase change, that is the phase modulation ability of the OASLM. However, V_{LC} on the LC layer of the OALCLV is difficult to measure directly in the experiment, so we have experimentally verified the effect of four write-light irradiances and four kinds of frequencies of V_{AC} on the phase change ability, and simulated these results using Equation (3), which are depicted in Figure 3.

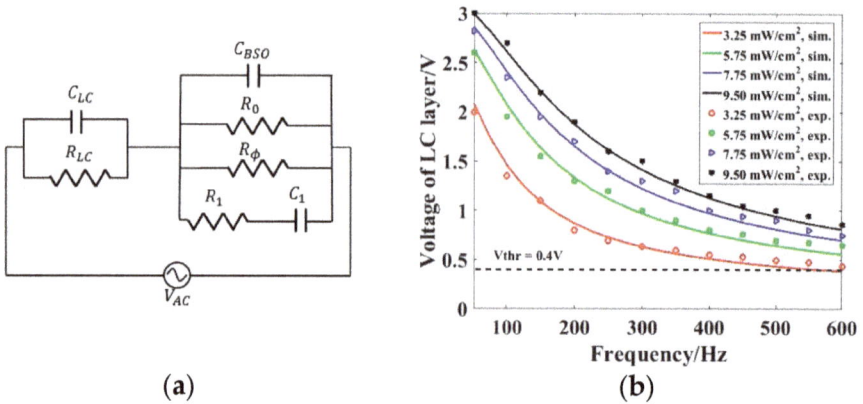

Figure 2. (**a**) Equivalent circuit of the OALCLV. (**b**) The voltage values on the LC layer of the OALCLV at the applied voltage of 3.5 V, experimental points and theoretical curves according to Equation (3), for different frequencies of the applied voltage.

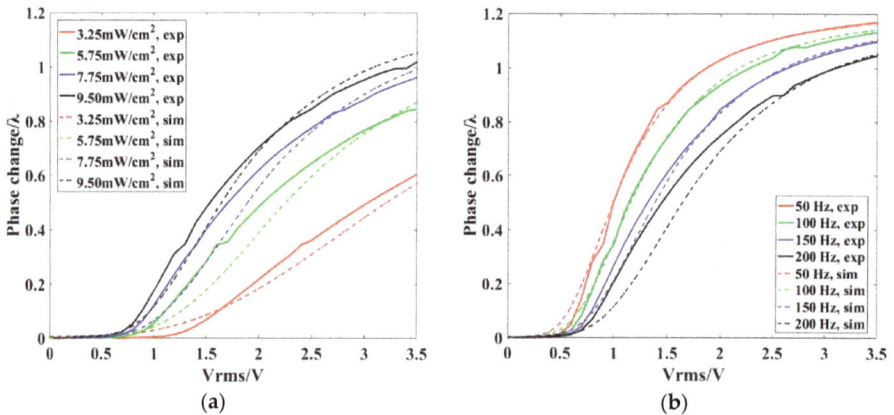

Figure 3. (**a**) Phase change varies with V_{AC} at 200 Hz when the OALCLV is illuminated by the different write light, the V_{rms} is the RMS value of the V_{AC}. (**b**) Phase change varies with V_{AC} when the OALCLV is at the different drive frequencies and the write-light irradiance is 9.6 mW/cm^2.

In Figure 3a, the solid lines show the relation between the phase change and V_{AC} at 200 Hz when the OASLM is subjected to four different write-light irradiances of 3.25 mW/cm^2, 5.75 mW/cm^2, 7.75 mW/cm^2, and 9.5 mW/cm^2. The dashed line shows the simulated phase-change relation deduced from the voltage response of the LC and Equation (3). The average error-rate between the simulated and experimental values is 3.5%. The figure illustrates that as the write-light irradiance decreases in the range of 9.5–3.25 mW/cm^2, the reduction of V_{LC} being approximately 1 V in Figure 2b, which results in a decrease in phase-change PV value of 0.4λ.

In Figure 3b, the solid lines show the relation between the phase change and V_{AC} at four different driving frequencies, namely 50 Hz, 100 Hz, 150 Hz, and 200 Hz when the OALCLV is uniformly illuminated with the write-light of irradiance 9.6 mW/cm^2. The dashed line shows the simulated phase-change relation. The average error-rate between the simulated and experimental values is 2%.

This figure illustrates that the reduction of the PV value is 0.1λ with the frequency increases from 50 Hz to 200 Hz, which corresponds to a decrease in V_{LC} of 1 V in Figure 2b.

In summary, the write-light irradiance and frequency of V_{AC} simultaneously affect the phase-change ability of the OALCLV in the form of change in V_{LC}, the voltage of the LC layer. At the same time, Equation (3) is a reliable formula to simulate V_{LC} and, when it is combined with the voltage response of the LC, the theoretical phase distribution can be accurately derived when the OALCLV is under different driving conditions.

The voltage across the LC layer affects the phase-modulation ability, which can be controlled by changing the write-light irradiance in the form of a grayscale image loaded on the LCOS in a practical application; therefore, we have experimentally verified the phase-modulation ability of the OASLM as a function of gray-level values when the LED is at different drive currents.

Figure 4a depicts the relationship between the phase change and grayscale values when the OASLM is under a V_{AC} of 14 V at 100 Hz and separately loaded these single-value grayscale images with values from 0 to 0.79 (in the normalized 0–255 grayscale range, 0–0.79 is the linear region of write-light irradiance as a function of grayscale value); meanwhile, the driving current of the LED changes from 200 mA to 800 mA. The dashed line shows the simulated phase-change relation deduced from the voltage response of the LC and Equation (3). The average error-rate between the simulated and experimental values is 2.2%. The figure shows that, with an increase in current, the decrease in the PV value, which is the phase-change ability, is 0.23λ, mainly because of the increase in the phase-change values corresponding to the gray level 0, implying that the corresponding V_{LC} is improved. In particular, we simulated the range of V_{LC}, which corresponds to 0–0.79 gray level when the OASLM is subjected to different driving conditions; the results are depicted in Figure 4b. The marked points correspond to the experimental values, which are basically consistent with the simulated values.

Figure 4b shows that, when the drive current changes from 200 mA to 800 mA, the write-light irradiance corresponding to the gray level of 0.79 increases from 3.25 mW/cm² to 9.50 mW/cm², the V_{LC} values at 100 Hz exceed the V_{sat} value, so all the phase-change values reach the maximum value of 0.97λ as depicted in Figure 4a; the increase in V_{LC} at 100 Hz is approximately 0.4 V as the write-light irradiance corresponding to the gray level of 0 increases from 0.131 mW/cm² to 0.388 mW/cm², and results in a decrease of the PV values of 0.23λ as depicted in Figure 4a. When the frequency is greater than 200 Hz, the values of the V_{LC} of the gray level 0 are all 0.5 V for different driving currents and do not change with increasing frequency. The current of the LED must be at least 400 mA at this time to ensure that the V_{LC} value for the gray level 0.79 exceeds the V_{sat} value, so we set the driving current at 400 mA, the corresponding dynamic range of the write-light irradiance is 0.235–5.75 mW/cm² to 0.388 mW/cm² when the grayscale value changes from 0 to 0.79 and the frequency is set at 200 Hz in practical applications.

The above analysis reveals that the range of the V_{LC} values corresponding to 0–0.79 gray level directly affects the phase-modulation capability of the OASLM. The ideal driving condition is that when the OASLM is loaded with 0–0.79 grayscale images, the range of V_{thr}–V_{sat} is at least contained in the variation range of V_{LC} to obtain theoretical phase-change capability. The equivalent circuit method combined with the voltage response curve of the LC can reliably simulate the phase-change relationship of OASLM and help us select the appropriate driving conditions, which will guide the development of the modulator with large phase-modulation ability in our future work.

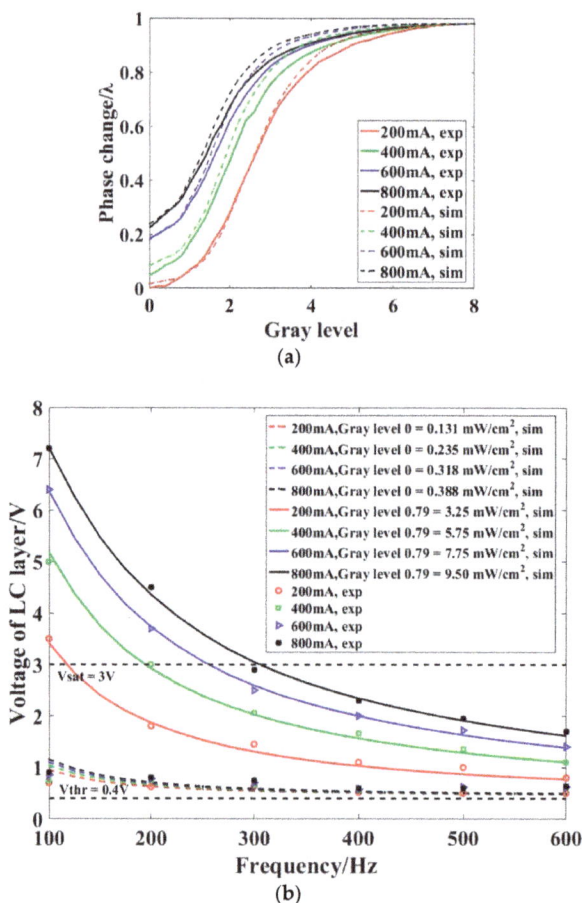

Figure 4. (**a**) Drive current of the write light affects the peak-to-valley (PV) value of the phase change when the OALCLV is at a V_{AC} value of 14 V at 100 Hz. (**b**) The voltage on the LC layer of the OALCLV as determined by the drive current of the write light.

3. An On-Line Monitor System for OASLM Modulation

The optical setup in Figure 5 is constructed for on-line monitoring of the phase modulation of the OASLM. A CCD with a spatial resolution of 512 pixels × 512 pixels is used to monitor the OASLM phase modulation and a WFS is used to detect the actual wave-front after the OASLM. In order to achieve phase-only modulation, the relative angle of the optical axis direction between the polarizer and the OALCLV is set to 0° for WFS measurement. The transmittance is measured by the CCD, when the relative angle of the optical axis direction between the polarizer and the OALCLV is set to 45°, and the polarizer and the analyzer are perpendicular to each other.

Figure 6a shows the transmittance when the OASLM is subjected to a current of write light of 400 mA and V_{AC} of 14 V at 200 Hz. The red curve in the legend is the transmittance of a single-value grayscale modulation, from which we obtain the phase-change function shown in Figure 6b with a PV value of 1λ. After 8-order polynomial fitting, this function is expressed as follows.

$$y = p_1x^8 + p_2x^7 + p_3x^6 + p_4x^5 + p_5x^4 + p_6x^3 + p_7x^2 + p_8x + p_9 \tag{4}$$

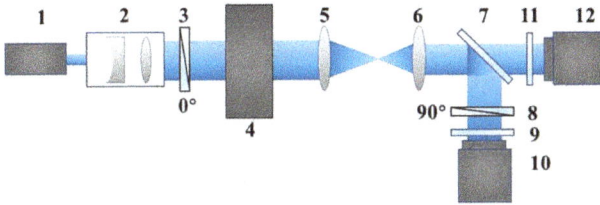

Figure 5. Optical setup used to measure the modulated wave-front. 1, 1053 nm continuous-wave Laser; 2, expander; 3, polarizer; 4, OASLM; 5, lens1, f1 = 150 mm; 6, lens2, f2 = 100 mm; 7, partially reflective mirror; 8, analyzer; 9, filter; 10, CCD; 11, filter; 12, wave-front sensor (WFS) (Physics Co., France, the spatial resolution is 29.6 µm).

The values of the fitting coefficients $p_1 - p_9$ are:

$$p_1 = 661.1715, \ p_2 = -2101.2010, \ p_3 = 2600.9500, \ p_4 = -1532.0300,$$

$$p_5 = 401.8500, \ p_6 = -28.0760, \ p_7 = 0.8321, \ p_8 = 0.2068, \ p_9 = -0.0002.$$

In order to further prove that the modulation results of the multi-value grayscale still satisfy the aforementioned relationship, we designed a gray-level ramp image and loaded it onto the OASLM to modulate the laser beam. The black curve in Figure 6a shows the corresponding one-dimensional (1D) transmittance distribution of the central row measured by the CCD, which has good consistency with the red curve.

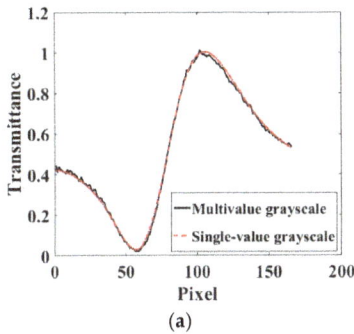

(a)

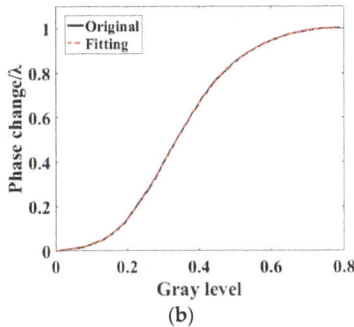

(b)

Figure 6. (**a**) Transmittance as a function of the grayscale. (**b**) Phase change as a function of the grayscale.

If we design a wave-front using Equation (4) and the actual modulation of the SLM loading the corresponding grayscale image is consistent with it, then we can assume that the transmittance measured by the CCD can monitor the phase-modulation result because the phase function in Equation (4) is directly derived from the transmittance. Figure 7a depicts a phase distribution designed with Equation (4), Figure 7b is the actual wave-front directly measured by the WFS and Figure 7c is a comparison of the 1D distribution of their central row. The modulated result is in good agreement with our designed wave-front.

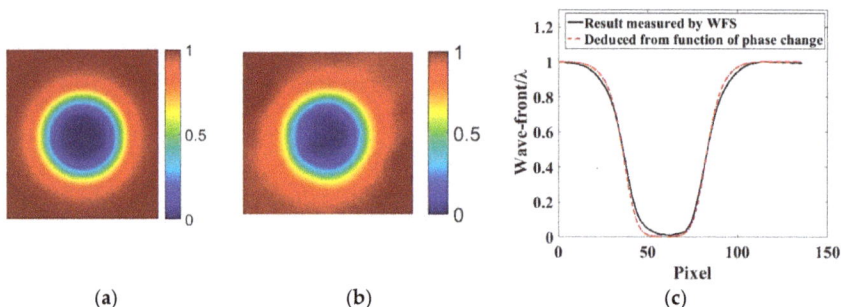

Figure 7. Comparison of the designed wave-front and the actual one. (**a**) Wave-front designed using Equation (4). (**b**)Actual wave-front measured by the WFS. (**c**) Comparison of the central row one-dimensional (1D) distribution.

Spatial resolution is also a key parameter of the monitoring system. In Figure 6a, the gray values corresponding to the maximum and minimum values are 0.25 and 0.45, respectively. Therefore, in order to demonstrate the accuracy of the system, we designed binary grayscale images with these two values, and the modulation results obtained by the CCD and WFS are compared with their theoretical values. Figure 8 shows modulation results of the OASLM loading Figure 9a (a binary image with a gray-value mutation). Figure 8a is a comparison between the 1D wave-front results obtained from the WFS and the theoretical distribution. Since the Hartmann wave-front sensor uses the wave-front slope and the wave-front variation to calculate the actual wave-front indirectly, here the modulation result of the binary image exceeds its measurement dynamic range, so the wave-front directly measured has a large recovery error and serious discrepancies compared to the theoretical distribution. Figure 8b is a comparison of the 1D transmittance between the result measured by the CCD and the result deduced from the transmittance gray-level relationship. The two results demonstrate good consistency. This implies that the monitoring system has a good response to the modulation of binary grayscale images.

Furthermore, we loaded a binary image with varying periods on our SLM to determine the spatial resolution of the system; the modulation result is shown in Figure 8c, from which we can see that the spatial resolution is at least 200 μm. In Figure 8d, the normalized MTF can be calculated from the relation $MTF = (T_{max} - T_{min})/(T_{max} + T_{min})$. From this we can also obtain the spatial resolution of 5 lp/mm (period of 200 μm, the value of the MTF is 0). At the same time, we can see that the 1 lp/mm (period of 1 mm, the value of the MTF is 1) is a division. When the value of the X-axis is larger than this, the MTF curve drops sharply to 0, so the minimum distinguishable contrast is 1 mm.

Figure 9b is a chessboard grayscale image with 70-pixel period in the horizontal direction and 128-pixel (the pixel-pitch of the write-light system is 20 μm) period in the vertical direction. The modulation results are shown in Figure 10. The black lines in (a) and (b) correspond to the transmittance measured by the CCD and are basically consistent with the red lines, the theoretical results of the transmittance gray-level relationship. The black lines in (c) and (d) are the 1D distributions of the wave-front measured by the WFS, which have significant differences compared to the theoretical values deduced from the phase-change function and represented by the red lines. Therefore, the monitoring system plays a very important role in the measurement of the modulation results on-line.

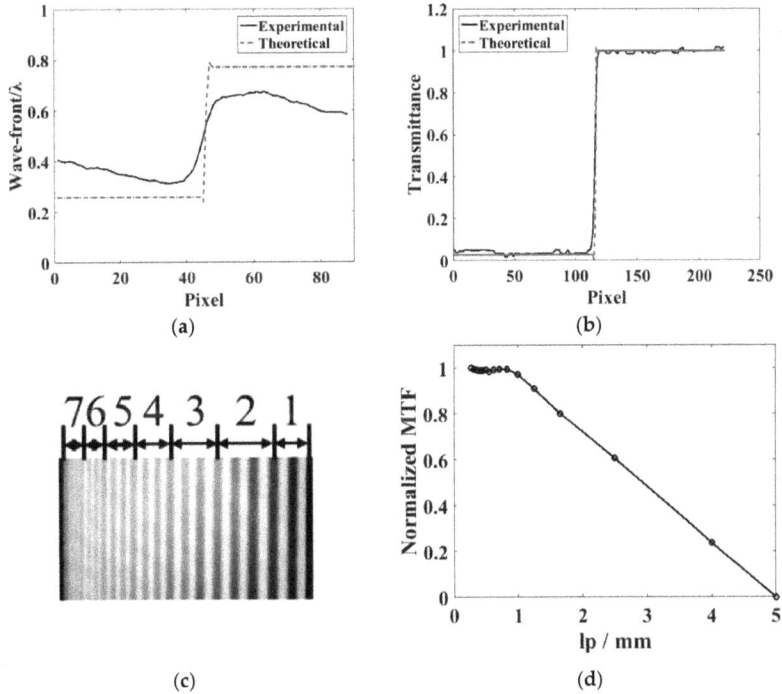

Figure 8. Modulation results of grayscale map with value mutation. (**a**) 1D wave-front. (**b**) 1D transmittance. (**c**) The modulated result obtained by CCD when the spatial-light modulator (SLM) is loaded with a binary image of varying variable periods. 1, 760 μm; 2, 520 μm; 3, 440 μm; 4, 320 μm; 5, 240 μm; 6, 200 μm; 7, 120 μm. (**d**) Modulation transfer function (MTF) of the OASLM as a function of spatial resolution.

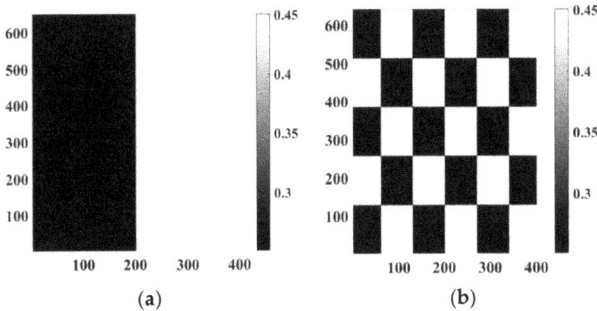

Figure 9. Binary images with gray values of 0.25 and 0.45. (**a**) 1D value change binary image. (**b**) Two-dimensional (2D) value change binary image.

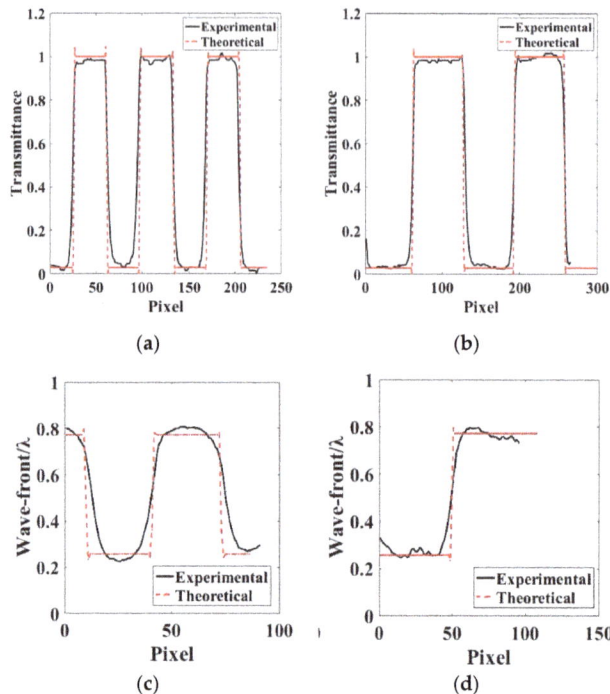

Figure 10. (**a**) 1D transmittance in the horizontal direction. (**b**) 1D transmittance in the vertical direction. (**c**) 1D wave-front in the horizontal direction. (**d**) 1D wave-front in the vertical direction.

4. Conclusions

We applied an equivalent circuit of the OALCLV to study the influence of driving conditions, including the frequency of drive voltage and irradiance of write light on the voltage of the LC layer; at the same time, we investigated the influence of driving conditions on the modulation ability of the phase-only OASLM. We were able to acquire a PV of phase change of approximately 1λ when the home-made OASLM was at the optimal driving condition. The detection system, including a CCD and WFS, was constructed to monitor the phase-control ability of the OASLM. From the MTF of the monitor system, we got a spatial resolution of 200 μm for modulation measurements and a minimum distinguishable contrast of 1 mm. We found that wave-front on-line modulation with feedback control could be obtained with the OASLM and the corresponding monitoring system.

Author Contributions: Conceptualization, W.F. and D.H.; Methodology, L.P.; Software, L.P.; Validation, W.F., D.H. and L.P.; Formal Analysis, L.P.; Investigation, L.P.; Resources, H.C.; Data Curation, D.H.; Writing-Original Draft Preparation, L.P.; Writing-Review & Editing, W.F. and D.H.; Visualization, W.F.; Supervision, X.L.; Project Administration, W.F.; Funding Acquisition, W.F.

Conflicts of Interest: The authors declare no conflict of interest.

References

1. Sala, F.A.; Karpierz, M.A. Modeling of molecular reorientation and beam propagation in chiral and non-chiral nematic liquid crystals. *Opt. Express* **2012**, *20*, 13923–13938. [CrossRef] [PubMed]
2. Chen, C.W.; Brigeman, A.N.; Ho, T.J.; Khoo, I.C. Normally transparent smart window based on electrically induced instability in dielectrically negative cholesteric liquid crystal. *Opt. Mater. Express* **2018**, *8*, 691–697. [CrossRef]

3. Saghaei, T.; Feiz, M.S.; Amjadi, A. Optical spatial phase retarder/modulator by a rotating freely suspended LC film. *Opt. Commun.* **2016**, *380*, 442–445. [CrossRef]

4. Bai, J.; Ge, M.; Wang, S.; Yang, Y.; Li, Y.; Chang, S. Characteristics of a liquid-crystal-filled composite lattice terahertz bandgap fiber. *Opt. Commun.* **2018**, *419*, 8–12. [CrossRef]

5. Huignard, J.P. Spatial light modulators and their applications. *J. Opt.* **1987**, *18*, 181–186. [CrossRef]

6. Aleksanyan, A.; Kravets, N.; Brasselet, E. Multiple-star system adaptive vortex coronagraphy using a liquid crystal light valve. *Phys. Rev. Lett.* **2017**, *118*, 203902. [CrossRef] [PubMed]

7. Peña, A.; Andersen, M.F. Complete polarization and phase control with a single spatial light modulator for the generation of complex light fields. *Laser Phys.* **2018**, *28*, 076201. [CrossRef]

8. Bortolozzo, U.; Dolfi, D.; Huignard, J.; Molin, P.S.; Peigné, A.; Residori, S. Phase modulation detection with liquid crystal devices. *Proc. SPIE* **2015**, *9378*, 93781S. [CrossRef]

9. Shrestha, P.K.; Chun, Y.T.; Chu, D. A high-resolution optically addressed spatial light modulator based on ZnO nanoparticles. *Light Sci. Appl.* **2015**, *4*, e259. [CrossRef]

10. Huang, D.; Fan, W.; Cheng, H.; Xia, G.; Pei, L.; Li, X.; Lin, Z. Wavefront control of laser beam using optically addressed liquid crystal modulator. *High Power Laser Sci. Eng.* **2018**, *6*, e20. [CrossRef]

11. Yamauchi, M.; Marquez, A.; Davis, J.A.; Franich, D.J. Interferometric phase measurements for polarization eigenvectors in twisted nematic liquid crystal spatial light modulators. *Opt. Commun.* **2000**, *181*, 1–6. [CrossRef]

12. Zhang, H.; Zhang, J.; Wu, L. Evaluation of phase–only liquid crystal spatial light modulator for phase modulation performance using a Twyman–Green interferometer. *Meas. Sci. Technol.* **2007**, *18*, 1724–1728. [CrossRef]

13. López-Téllez, J.M.; Bruce, N.C.; Rodríguez-Herrera, O.G. Characterization of optical polarization properties for liquid crystal-based retarders. *Appl. Opt.* **2016**, *55*, 6025–6033. [CrossRef] [PubMed]

14. Aldrich, R.E.; Hou, S.L.; Harvill, M.L. Electrical and Optical Properties of $Bi_{12}SiO_{20}$. *J. Appl. Phys.* **1971**, *42*, 493. [CrossRef]

15. Lu, K.; Saleh, B.E.A. Complex amplitude reflectance of the liquid crystal light valve. *Appl. Opt.* **1991**, *30*, 2354–2362. [CrossRef] [PubMed]

16. Cao, Z.; Xuan, L.; Hu, L.; Liu, Y.; Mu, Q.; Li, D. Investigation of optical testing with a phase-only liquid crystal spatial light modulator. *Opt. Express* **2005**, *13*, 1059–1065. [CrossRef] [PubMed]

17. López-Téllez, J.M.; Bruce, N.C.; Delgado-Aguillón, J.; Garduño-Mejía, J.; Avendaño-Alejo, M. Experimental method to characterize the retardance function of optical variable retarders. *J. Phys.* **2015**, *83*, 143–149. [CrossRef]

18. López-Téllez, J.M.; Bruce, N.C. Stokes polarimetry using analysis of the nonlinear voltage-retardance relationship for liquid-crystal variable retarders. *Rev. Sci. Instrum.* **2014**, *85*, 033104. [CrossRef] [PubMed]

19. Huang, D.; Fan, W.; Li, X.; Lin, Z. Performance of an optically addressed liquid crystal light valve and its application in optics damage protection. *Chin. Opt. Lett.* **2013**, *11*, 072301. [CrossRef]

20. Aubourg, P.; Huignard, J.P.; Hareng, M.; Mullen, R.A. Liquid crystal light valve using bulk monocrystalline $Bi_{12}SiO_{20}$ as the photoconductive material. *Appl. Opt.* **1982**, *21*, 3706–3712. [CrossRef] [PubMed]

![applied sciences logo] *applied* *sciences*

MDPI

Review

Liquid-Crystal-on-Silicon for Augmented Reality Displays

Yuge Huang [1], Engle Liao [2], Ran Chen [1,3] and Shin-Tson Wu [1,*]

[1] College of Optics and Photonics, University of Central Florida, Orlando, FL 32816, USA;
    y.huang@knights.ucf.edu (Y.H.); tradchenr@knights.ucf.edu (R.C.)
[2] Snail Innovation Institute, San Jose, CA 95112, USA; englel@usasii.com
[3] Key Laboratory of Applied Surface and Colloid Chemistry, School of Materials Science and Engineering,
    Shaanxi Normal University, Xi'an 710119, China
*   Correspondence: swu@creol.ucf.edu; Tel.: +1-407-823-4763

Received: 30 October 2018; Accepted: 20 November 2018; Published: 23 November 2018

Abstract: In this paper, we review liquid-crystal-on-silicon (LCoS) technology and focus on its new application in emerging augmented reality (AR) displays. In the first part, the LCoS working principles of three commonly adopted LC modes—vertical alignment and twist nematic for amplitude modulation, and homogeneous alignment for phase modulation—are introduced and their pros and cons evaluated. In the second part, the fringing field effect is analyzed, and a novel pretilt angle patterning method for suppressing the effect is presented. Moreover, we illustrate how to integrate the LCoS panel in an AR display system. Both currently available intensity modulators and under-developing holographic displays are covered, with special emphases on achieving high image quality, such as a fast response time and high-resolution. The rapidly increasing application of LCoS in AR head-mounted displays and head-up displays is foreseeable.

Keywords: liquid-crystal-on-silicon; fringing field effect; augmented reality displays; head-mounted displays; head-up displays

1. Introduction

Dating from the 1970s, Hughes research labs creatively combined liquid crystal (LC) with semiconductor substrates, enabling optically addressed liquid crystal light valves [1,2], which paved the way for the invention of liquid-crystal-on-silicon (LCoS) display technology [3–5]. Through decades of extensive efforts, LCoS technology has proven its success in a wide range of fields like displays [6–13], adaptive optics [14,15], metrology [16,17], telecommunications [18], quantum physics [19,20], etc. The emergence of LCoS is jointly contributed by the application of liquid crystal (LC) electro-optic response behaviors and the development of the silicon complementary metal oxide semiconductor (CMOS) backplane technique. The structure of a typical reflective LCoS panel is shown in Figure 1: an LC layer is sandwiched between an indium tin oxide (ITO)-coated glass and a CMOS backplane. Two thin polyimide (PI) alignment layers determine the LC alignment direction. The Aluminum (Al) electrodes on the Silicon CMOS backplane are pixelated to provide independent voltage control. When an incident linearly polarized light traversing through the LC layer twice, its polarization state or phase is modulated depending on the LC alignment mode.

Recently, LCoS has been widely implemented in augmented reality (AR) displays because of its high luminance (>30,000 nits), compactness, high-resolution density (>4000 pixels per inch), and potential of holography generation [12,21–23]. However, the cutting edge LCoS still needs improvement to meet the stringent AR requirements. In this paper, we review the LCoS technology intended for AR displays. In Section 2, we briefly describe the basic light modulation principles of LCoS. In Section 3, we address the fringing field effect issue impairing the LCoS display performance, and propose a new device structure to mitigate this problem. High-resolution and small-pitch LCoS panels are especially desirable for augmented reality displays, but how to suppress the fringing field effect remains a technical challenge. In the last section, the LCoS-integrated AR head-mounted display (HMD) and head-up display (HUD) systems are demonstrated. The key parameters determining the display performance are discussed in detail.

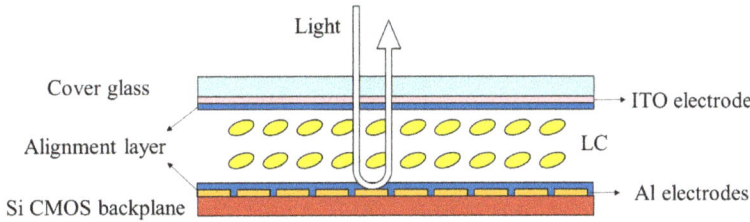

Figure 1. The schematic of an LCoS structure.

2. LCoS Working Principles

2.1. Birefringence Effect of Liquid Crystals

Figure 2 illustrates the working mechanism of an LC phase retarder. Here, we assume the LC molecules are uniformly aligned along the x-axis with a tilt angle θ (Figure 2a). The uniaxial LC exhibits an extraordinary refractive index n_e along its optical axis and an ordinary refractive index n_o along the two orthogonal axes. Thus, the corresponding birefringence is defined as $\Delta n = n_e - n_o$. In Figure 2b, when an incident light passes through the LC layer along the z-axis, the x-polarized light experiences a tilt-angle-dependent refractive index $n_x(\theta)$, while the y-polarized light experiences a constant index $n_y = n_o$. The accumulated phase retardation between the two linear polarizations is

$$\Gamma = [n_x - n_y] \cdot \frac{2\pi}{\lambda} \cdot d. \tag{1}$$

Here d is the cell gap (LC layer thickness), and $n_x(\theta)$ follows

$$\frac{1}{n_x(\theta)^2} = \frac{\cos^2\theta}{n_e{}^2} + \frac{\sin^2\theta}{n_o{}^2}. \tag{2}$$

In a reflective display, the incident light passes through the LC layer twice. As a result, the effective cell gap is $2d$ and Γ is doubled.

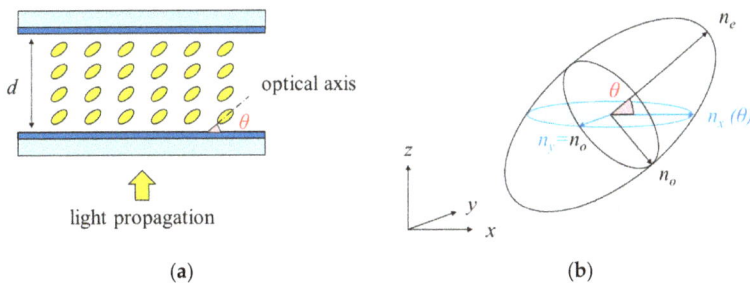

Figure 2. (a) The schematic structure of an LC phase retarder. (b) Refractive index ellipse of a uniaxial liquid crystal.

When an external voltage is applied, the LC molecules (with $\Delta\varepsilon > 0$) would reorient toward the electric field direction which, in turn, modifies the phase and polarization of the outgoing light. Depending on the surface alignment and the LC material properties, three LC modes are widely adopted: vertical alignment (VA) and mixed-mode twist nematic (MTN) mainly for intensity modulation, as well as the homogeneous alignment for phase modulation.

2.2. Vertical Alignment Mode

In the VA mode [24], the initial LC directors are nearly perpendicular to the substrates (pretilt angle $\alpha \sim 88°$), as Figure 3 shows. For intensity modulation, the VA cell is sandwiched between two crossed polarizers, whose absorption axes are $\pm 45°$ with respect to the x-axis. In the voltage-off state ($V = 0$), the incident light propagating along the z-direction experiences a negligible phase retardation. That means that the linearly polarized incident light remains unchanged and cannot pass through the crossed analyzer. As the applied voltage exceeds a threshold, the LC molecules with a negative dielectric anisotropy ($\Delta\varepsilon < 0$) would be reoriented perpendicular to the longitudinal electric field. Because the surface anchoring force decreases as the distance from the alignment layer increases, the LC molecules keep the pretilt angle near the substrates and get fully reoriented in the central layer (Figure 3b). A pretilt angle $\alpha \neq 90°$ gives a preference of x-z plane reorientation. Balancing between the competitive surface anchoring force, electric force and elastic restoring force, the LC tilt angle can change from ~88° towards 0° as the voltage increases. The associated phase retardation can be tuned from 0 to π, thus, the output light intensity can be progressively changed from dark to bright. The VA mode produces an excellent dark state and high contrast ratio (CR > 5000:1) regardless of the cell gap, wavelength, and operation temperature. However, the fringing field effect on the VA mode is more severe, which will be addressed in detail in the next section.

In principle, a VA cell can also be used for phase-only modulation, if the incident light polarization is parallel to the LC directors (in the voltage-on state). However, the negative-$\Delta\varepsilon$ LC materials usually exhibit a higher viscosity and a smaller $|\Delta\varepsilon|$ than the corresponding positive LCs because the dipoles are in the lateral positions. As a result, homogeneous cells using a positive-$\Delta\varepsilon$ LC are preferred for phase-only spatial light modulators.

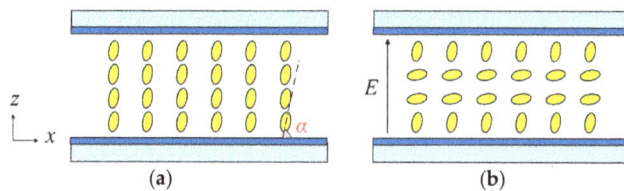

Figure 3. The LC director orientations in VA mode: (**a**) Voltage-off state; (**b**) Voltage-on state.

2.3. Mixed-Mode Twist Nematic (MTN) Mode

In a twisted-nematic (TN) mode [25,26], the LCs form twisted structures by the surface alignment and the doped chiral agent. For example, in a 90° TN cell, the top and the bottom LCs are aligned orthogonally with a twist angle $\phi = 90°$, as illustrated in Figure 4a. Such a molecular alignment can be achieved by mechanical rubbing on polyimide (PI) [15,27] or by the photo-alignment technique [28,29]. For intensity modulation, the LC cell is sandwiched between two crossed polarizers. The polarization axis of the front polarizer has an in-plane angle β with respect to the front LC alignment. When $\beta = 0°$, the polarization rotation effect takes place, which guides the incident x-linearly polarized light to y-linear polarization. The Gooch–Tarry first minimum condition $d\Delta n/\lambda = \sqrt{3}/2$ [30] determines the thinnest cell gap for a given LC birefringence to achieve maximum transmittance. When a longitudinal electric field is applied, the LC molecules ($\Delta\varepsilon > 0$) are reoriented toward the vertical direction and a dark state is produced (Figure 4b). TN cells work well as a normally-white intensity modulator. It can

Appl. Sci. **2018**, *8*, 2366

also be used for phase-only modulation [31], but the voltage should be controlled below the optical threshold [32], leading to a slow response time [33]. For intensity modulation, the response time of the TN mode and VA mode could be comparable, depending on the employed LC material and the cell gap. A major drawback of TN mode is the compromised contrast ratio (CR ~1000:1).

In a reflective LCoS, the doubled optical path leads to a smaller $d\Delta n$ requirement. However, when $\beta = 0°$ the normalized optical efficiency is only 70% [6]. Two methods can be used to boost the optical efficiency: (1) to increase β to about 20°, and (2) to decrease the twist angle to 80° or 70° [34]. At $\beta \approx 20°$, both the polarization rotation effect and birefringence effect coexist. This is called the mixed-mode twisted nematic (MTN) mode [6]. The optical efficiency increases to ~88%. Decreasing the twist angle can enhance the optical efficiency to nearly 100%, but the contrast ratio is decreased dramatically. Therefore, 90° MTN a favored choice, especially for field sequential color (FSC) displays, because of its merits in high contrast ratio, low operation voltage and fast response time. By removing the spatial color filters, FSC operation could triple the optical efficiency and resolution density.

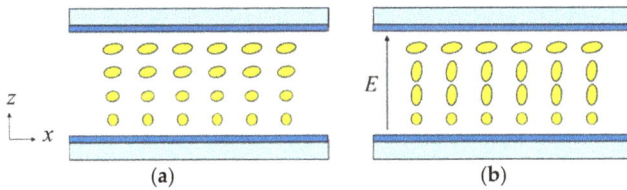

Figure 4. The LC director orientations in the 90° TN mode: (**a**) Voltage-off state; (**b**) Voltage-on state.

2.4. Homogeneous Alignment

In a homogeneous (HG) cell (Figure 5), the LC alignment is parallel to the x-axis with a small pretilt angle ($\alpha \sim 2°$) [29]. Using a positive $\Delta\varepsilon$ LC material, its electro-optic properties are opposite to the VA mode: phase retardation is maximized at the voltage-off state and decreases as the applied voltage increases. Due to the narrow viewing angle and the staggered RGB dark state, the HG cell is rarely used for broadband intensity modulation but is a favored choice for a reflective phase modulator. The viewing angle can be enlarged by a compensation film and by the mirror image effect in reflective LCoS [35]. The unmatched RGB phase retardation is not a big issue for phase modulation. The major advantage of the HG mode is its fast response time. To achieve the 2π phase range, the LC layer is usually twice thicker than its counterpart intensity modulator, as a result, its response time (τ is proportional to d^2) would be 4× slower [33]. Thanks to the wide selection of positive LC materials, high $\Delta\varepsilon$ for low operation voltage and low rotational viscosity γ_1 for fast response time can be obtained more easily than those negative-$\Delta\varepsilon$ LCs for the VA mode [36].

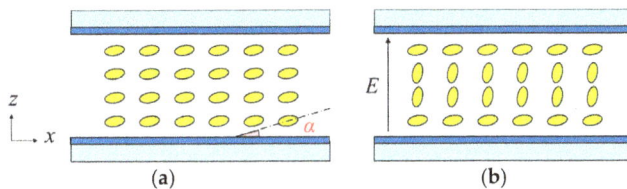

Figure 5. The LC director orientations in the HG mode: (**a**) Voltage-off state; (**b**) Voltage-on state.

2.5. LC Material Optimization Strategies

In electro-optic modulation, a general requirement applies to all of the above-mentioned modes: $d\Delta n/\lambda$ equals a certain value. Regarding electrical properties, the CMOS circuitry limits the operation voltage to <6 V for small-pitch panels. Such a low voltage can be achieved by increasing the dielectric anisotropy of LC materials, which could be obtained by applying lateral and terminal polar groups (such as F, CN, NCS, CF_3 or OCF_3) [37–39], polar CF_2O bridges [40], and polar heterocycles [41,42]. However, these components often lead to a compromised nematic temperature range, lower solubility and higher viscosity γ_1 [38]. Breaking this trade-off, the difluoromethoxy bridge (-CF_2O) [43] and isothiocyanato group (-NCS) [44,45] could improve $\Delta\varepsilon$ while retaining a reasonably low γ_1. Whereas the former breaks the molecular π-π conjugation resulting in a low Δn, and the latter gives rise to the annoying charge accumulation effect, which degrades the voltage holding ratio. Here comes the second consideration on electrical properties—the resistivity which determines the voltage holding ratio. Although LCoS devices demand a lower LC resistivity (>10^{12} Ω·cm) than the active matrix displays (>10^{13} Ω·cm) [46,47], it still limits the LC component selection. Compounds with a large $\Delta\varepsilon$ would require multiple dipoles, which may lead to a compromised γ_1.

From the aspect of device configuration, a thinner cell gap helps gain fast response time with the adoption of a slightly higher Δn LC. To be noticed, the toughest color to achieve the 2π phase modulation is red because of the longest wavelength and the lowest birefringence. Various terphenyls and tolanes have been developed for high Δn LCs [48,49]. Although many precise modifications of the molecular chemical structures have been mentioned [50–52], such as shifting the location of the mesomorphic core, introducing the lateral substituents into different positions and the various terminals, the high γ_1 of terphenyls and the poor UV stability of tolanes remains unsolved [53]. Those are the properties not intended to be compromised. Additionally, the voltage shielding effect by alignment layer reduces the effective voltage applied to the LC layer in thin cells [54,55]. In summary, a delicate balance should be reached between high birefringence, high dielectric anisotropy, low melting point, low viscosity, high resistivity, and good UV stability according to the device configuration. The state-of-the-art LC material development reaches a submillisecond response time for VA [56–58] and MTN [46,59] LCoS at 40 °C, and a ~2 ms response time for 2π phase modulation [46,47]. More than that, these optimization strategies and the developed materials could be applied to other LC phase modulation-based devices such as tunable lenses, scanners, tunable filters, and optically addressed spatial light modulators [60–64], which may be incorporated in AR systems in other ways.

3. Fringing Field Effect and Novel Solution

To optimize the optical performance of LCoS systematically, several research groups have devoted efforts in device modeling [65–67]. However, until recently, the fringing field effect (FFE) still remains the major issue to be tackled with [68–73]. In order to display greyscale images, different pixels could have different voltages. The unequal voltages on adjacent pixels give rises to FFE. As Figure 6 depicts, when a bright pixel is adjacent to two dark pixels, a dark split region may appear in the bright pixel area, and light leakage arises at the edge of the dark pixel, leading to display luminance loss and reduced contrast ratio. The severity of FFE depends on the display mode and the device configuration. Generally speaking, VA has the strongest FFE, resulting in a ~30% luminance loss; 90° MTN is the least sensitive to FFE; while the HG mode is a favored choice for phase modulation. Figure 6 shows the FFE of VA mode with different pixel sizes and cell gap. Because of the small inter-pixel gap, LCs on the pixel edge experience strong horizontal electric field, giving rise to phase distortion. This phenomenon is not obvious in direct-view transmissive LCDs because the inter-pixel region is covered by a large-area black matrix (>5 μm). The asymmetry of phase retardation/reflection is determined by the alignment direction of the pretilt angle. In Figure 6a, the same resolution density is compared, which is determined by pixel pitch p. With a thinner cell gap d, the FFE-affected region is compressed to a narrower edge. Applying this thin cell strategy [69,70], the cell gap in VA LCoS is optimized to $d < 2$ μm [73].

FFE is especially pronounced in small-pitch LCoS devices [69], which is highly desirable in augmented reality applications. Because of the reduced pixel pitch and cell gap (p/d) ratio, the unwanted horizontal electric field is strengthened [70]. This can be seen from Figure 6b: when the resolution density is doubled (p changes from 16 μm to 8 μm), the absolute FFE width w is relatively stable, while the ratio of the FFE-affected region increases dramatically. Several methods have been proposed to reduce FFE. In 2002, Fan-Chiang et al. explored how sloped electrodes help reduce FFE with quantitative exploration [68]. An optimal electrode slope of 1 was demonstrated for 80° MTN cell and 45° TN cell, while the VA and HG modes call for further improvement. In the same year, a finger-on-plane type common electrode was designed to eliminate the pixel-splitting phenomenon [74]. Whereas the diffraction effect and relatively low optical efficiency hinder its widespread applications [69]. In 2005, Gu et al. minimized FFE by a double-sided electrode design, where the requirements of dual TFTs and pixel registration brings manufacturing challenges [75]. Later, Fan-Chiang et al. proposed to adopt circularly polarized light in the VA mode for FFE suppression [69]. Though an impressively high optical efficiency and fast dynamic response were demonstrated, hardly can this method be integrated into the preferred normally black reflective VA mode. In other words, the contrast ratio is compromised. In 2015, Li et al. adopted non-rectangular pixel electrodes for pixel edge FFE compensation [76]. This design works well on large pixels while the degree of structural modification is constrained on small-pitch configurations. To satisfy the urgent need of FFE reduction on small-pitch LCoS devices, here, we propose a novel design of patterned pretilt angle control [77–79]. The formed inhomogeneous LC director distribution could oppose the FFE induced by neighboring pixels.

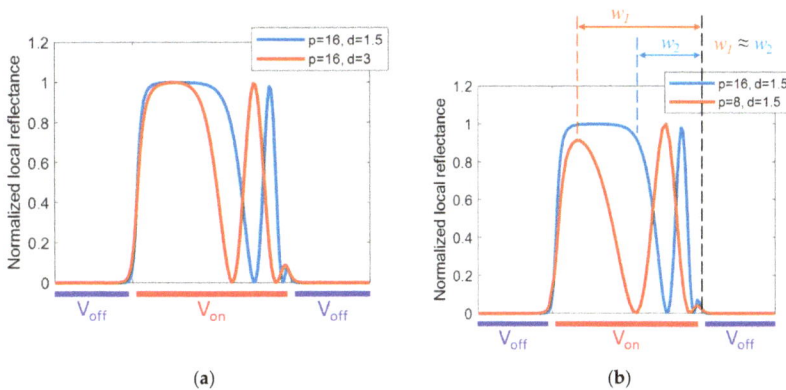

Figure 6. The fringing field effect (FFE) of a VA cell: (**a**) cell gap effect; (**b**) pixel pitch effect. p, d, and w stand for pixel pitch, cell gap, and FFE border width, respectively (unit: μm).

3.1. Pretilt Angle Pattern Determining Method

The basic concept of our design is to find an optimized pretilt angle pattern for minimizing FFE. The initial step is to identify the phase retardation profile as a function of location in the pixel region. Since FFE is most pronounced when the largest voltage difference occurs between adjacent pixel electrodes, two worst scenarios are used in our evaluation and optimization. These are (1) maximum operation voltage applied to the central pixel electrode while minimum operation voltage applied to the adjacent pixel electrodes; and (2) minimum operation voltage for the central pixel while the maximum voltage for the neighboring ones.

After basic evaluation, there are two determining methods to optimize the pretilt angle pattern. The first is to control the phase retardation error over the whole pixel region within a predetermined percentage level. As a general determining approach, the lack of a convergent solution makes the optimization complicated, and the iterative algorism results in heavy computation loads. Therefore, here we introduce the second determining method. As shown in Figure 6b, based on the initially identified phase retardation profile, we divide the pixel region into two zones with an FFE border width w. From Figure 7, we can see that the pretilt angle of the inner zone and outer zone is optimized to be α_i and α_o, respectively. The target of the optimization is to reduce the overall phase retardation error. With this simpler determining approach, pretilt angle patterns with an FFE-opposing feature can be designed.

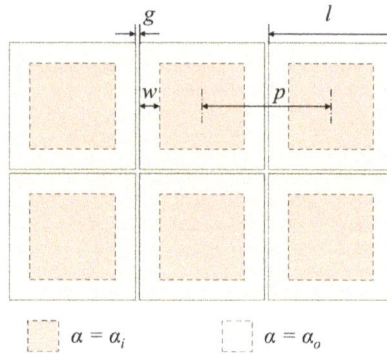

Figure 7. The LCoS pixel layout and optimization parameters: pixel length l, pixel gap g, pixel pitch p, FFE border width w, inner zone pretilt angle α_i, and inner zone pretilt angle α_o.

3.2. Exemplary Optimization Results

Here we use an exemplary optimized structure to demonstrate the performance of the proposed design. The basic HG LCoS parameters are set as follows: cell gap d = 1.5 μm, pixel length l = 6.2 μm and pixel gap g = 0.2 μm, which corresponds to pitch length $p = l + g$ = 6.4 μm. The LC parameters are Δn = 0.16 at 550 nm, $\Delta\varepsilon$ = 3.1, γ_1 = 118 mPa·s, and (splay, twist, bend) elastic constants (K_{11}, K_{22}, K_{33}) = (13.2, 6.5, 18.3) pN at 25 °C (from Merck). As illustrated in Figure 8a, by simulating the phase retardation profile with an even pretilt angle, the FFE-affected outer zone and the unaffected inner zone are differentiated with w = 1 μm. Applying the second determining method, the optimized inner zone pretilt angle α_i = 88° and outer zone pretilt angle α_o = 85° are obtained. Figure 8b shows the results with our modified pretilt angle pattern design: the electric field distortion at pixel edge is much suppressed and the FFE is effectively reduced.

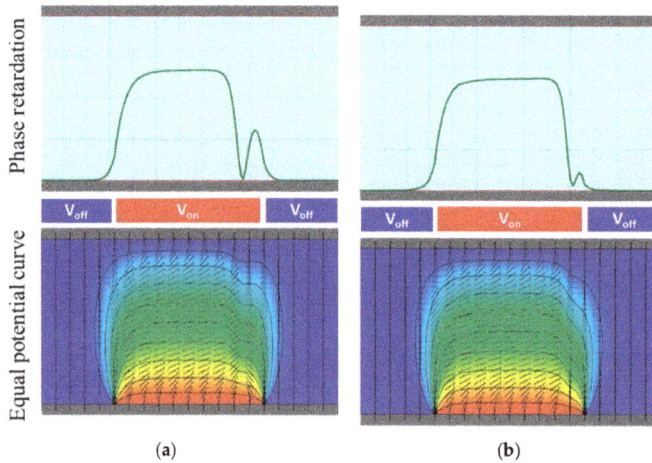

Figure 8. The pixel location dependent phase retardation curve and equal potential curve with (**a**) even pretilt angle distribution and (**b**) specially-designed uneven pretilt angle distribution.

One potential concern of our design is the phase retardation uniformity when an equal voltage is applied to adjacent pixel electrodes. Therefore, we plot the phase retardation curves with our special pretilt angle design. From Figure 9a we can see, in the large pretilt angle region ($\alpha > 80°$), that neither phase retardation nor reflectance is sensitive to the pretilt angle. In Figure 9b, generally flat spatial profiles are obtained in our design. The spatial fluctuation at all-pixels-dark state is only $\Delta\Gamma = \Gamma~(\alpha = 85°) - \Gamma~(\alpha = 88°) = 0.01\pi$ for phase retardation and R $(\alpha = 85°)$ − R $(\alpha = 88°) = 0.03\%$ for reflectance, suggesting that a high contrast ratio could be maintained. At all-pixels-bright state, the reflectance fluctuation is within 0.002%. In other words, the side effect of our proposed design is negligible compared to the gain in FFE reduction.

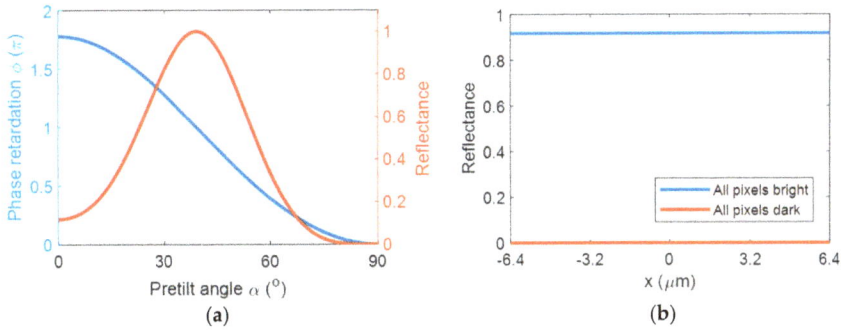

Figure 9. (a) The pretilt angle dependent phase retardations. (b) Spatial phase retardation uniformity at voltage-off state.

3.3. Fabrication Method

The pretilt angle pattern can be realized by a photoalignment layer, which provides inhomogeneous anchoring energy distribution. Two methods have been proposed to realize the spatial variant pretilt angle control by Kwok's group. In 2011, the stack of a horizontal alignment polymer on a vertical alignment layer was reported [80]. By controlling the dosage of UV light exposure, the degree of polymerization can be adjusted spatially. After rinsing away the unexposed area, the ratio of polymerized homogeneous alignment domain and exposed vertical alignment domain coordinately determine the local LC pretilt angle. In 2013, Fan et al. reported a new alignment material which enables direct UV patterning on a pretilt angle [81]. Based on the above-mentioned methods, a two-step manufacturing process can generate FFE-opposing features (Figure 10): (1) A uniform pretilt angle α_1 is produced on the polyimide (PI) alignment layer on ITO glass by uniform UV irradiation; (2) A second photomask for selective UV exposure is added to differentiate the pretilt angle α_i in inner zone and α_o in outer zone. If the alignment layer is replaced by silicon oxide [82], an e-beam treatment can be adopted for the patterning.

For infrared spatial light modulators, the patterned pretilt angle and anchoring energy can also be achieved on the polymer-stabilized liquid crystal (PSLC) [83,84]. First, an LC-monomer mixed precursor is prepared and filled in the LCoS panel. Second, a photomask with gradient optical density is required to generate spatially variant UV light intensity. The UV light irradiated on the LC layer is more intense in the center region than at the pixel edge, therefore, in the center, it generates a higher degree of polymerization. Through this spatially variant polymerization process, an inhomogeneous pretilt angle and anchoring energy distribution can be formed, as Figure 11 illustrates.

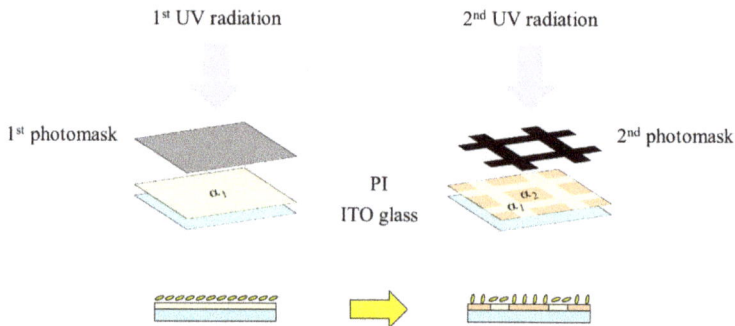

Figure 10. The proposed fabrication procedure of an LCoS with FFE-opposing features.

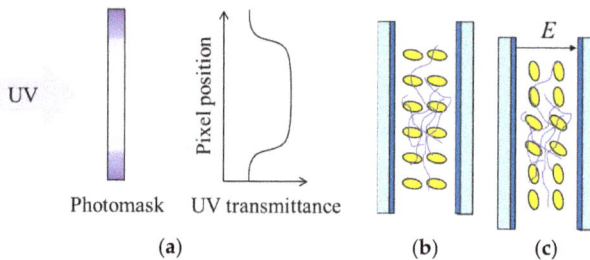

Figure 11. The proposed fabrication method of a PSLC LCoS with FFE-opposing features: (**a**) UV transmittance control by a gradient optical density photomask. (**b**) HG cell with inhomogeneous polymerization. (**c**) Spatially variant LC response to an external electric field.

4. LCoS for Augmented Reality Displays

4.1. Augmented Reality Head-Mounted Displays

So far, LCoS is the dominant solution for augmented reality (AR) head-mounted display (HMD) because of its compact size (<1.5 inches), high optical efficiency (>90%), high fill factor (>90%), and high resolution (4K2K achieved). In 2013, Kress et al. reviewed the HMD optical systems focusing on the combiner designs [85]. Here, we just use Figure 12 to schematically show how an LCoS functions in an AR-HMD optical system (Google Glass's design [86]). A polarizing beam splitter (PBS) is engaged as crossed polarizers for the reflective display system. When the light emitted from an LED source enters the PBS (purple line), the s-polarized light (red line) is reflected away so that only the p-polarized light (blue line) can reach the LCoS panel, indicating that only half of the incident light could be modulated. To enhance the display brightness, polarization conversion optical elements have been developed [87] and can be inserted between the LED source and the PBS together with collimation lenses. Passing through the LCoS, the unmodulated p-polarized light (thin blue line) transmits back to the LED source, while the π-phase retarded s-polarized light (red) gets reflected by the PBS and directed to human eyes by a partially-reflective mirror (PRM) and a focusing mirror (FM). As the ambient light transmits PRM as well (green line), an AR experience with an LCoS-generated display and real world can be achieved.

Current 3D displays are mainly based on binocular disparity: two eyes receive different images representing slightly different viewing points. However, the mismatch of other focus cues gives rise to visual discomfort [88,89] and distorted depth [90–92]. As a promising one-stop solution to provide correct focus cues with aberration correction [93], phase-only LCoS-based AR-HMD has been demonstrated [12,13,21]. Different from the intensity modulation configuration, the incident linearly polarized light on phase-only LCoS should be parallel to the LC directors. Therefore,

PBS could be eliminated in phase-only AR systems, enabling an eyeglasses size form factor [12]. In contrast to the hardware lightness, the computational burden on phase encoding is heavy. In the past decades, extensive efforts have been devoted to developing high-quality real-time holography algorithms [94–98]. Though prototypes have been demonstrated [12], the commercialization of high-quality holographic AR-HMD may still need time.

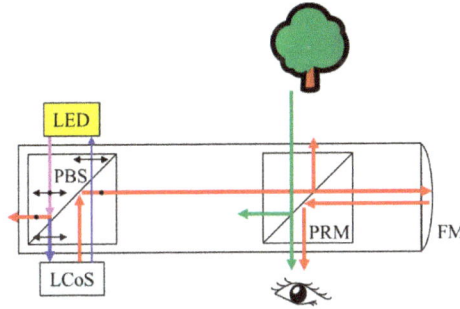

Figure 12. The schematic of LCoS based AR-HMD optical system. PBS: polarizing beam splitter; PRM: partially-reflective mirror; FM: focusing mirror; blue lines: p-polarized light; red lines: s-polarized light; purple lines: p and s polarized light.

4.1.1. Response Time

The LCoS response time greatly affects the AR viewing experience. To avoid the flickering effect [99], a refresh rate of 60 Hz is the minimum requirement, while >144 Hz is highly favorable. However, the relatively slow LC response time could impair grey level accuracy at high frame rates, as Figure 13 illustrates. Taking 120 Hz as an example, the frame time is 8.3 ms. The green curve shows the applied voltage at the two consecutive frames. To switch from dark to bright state, the voltage is raised from the minimum to the maximum value, as the green curve denotes. The LC molecules respond gradually so that the reflectance increases from 0 to 100% during the first frame (0~8.3 ms) (blue curve). As a result, the *average* reflectance in the first frame is much lower than the expected 100%, as the red dashed lines mark. Similarly, if the voltage is released at the second frame, it only generates a relatively poor dark state. The relatively slow response time (tens of milliseconds) of conventional LC materials cannot satisfy the >60 Hz refresh requirement (>180 Hz color field rate). To overcome the slow response issue in the above-mentioned analog driving scheme, a digital driving scheme, say, pulse width modulation (PWM), is widely adopted in an LCoS intensity modulator [100]. By applying binary voltages with various pulse widths within one frame, the LC molecules respond to the root-mean-square (RMS) voltages and human eyes perceive the average reflectance. Although digital driving lays less burden on LC response time and provides better grey level accuracy, stronger FFE may appear with small grey-level changes. In 2000, Worley et al. designed a bit-splitting method to overcome the FFE in digital driving by sub-frame rate increasing [101]. The LC response time in digital driving scheme needs to be carefully selected: it should be less than the frame time, but longer than the pulse width to avoid a perceivable light intensity fluctuation [102]. In theory, digital driving is not suitable for phase-only LCoS because phase information cannot be averaged with an RMS value. In practice, phase flickering has been suppressed to an acceptable level [103]. Jointly with its merits of fast response time, high repeatability and good phase vs. gray level linearity, digital driving is widely adopted in commercial phase-only LCoS devices [104]. In general, fast-response LC materials help improve the performance of both analog-driven and digital-driven LCoS devices. Therefore, extensive efforts have been made to obtain submillisecond response LCs [46,57–59].

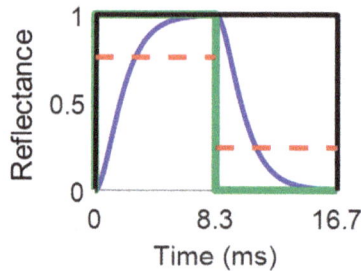

Figure 13. An illustration of the slow LC response time-impaired gray level accuracy.

4.1.2. Resolution

To cover the ~100° field of view (FoV) with human eye acuity 1 arcminute, ideally, a 6K6K resolution is required for each eye [105]. As a reference, the resolution of most commercially available products is ≤4K2K, which is still insufficient. Increasing the hardware resolution brings several challenges: (1) If the panel size remains the same, then increasing resolution density implies a smaller pixel size, which would result in a stronger FFE [69]. The limit of inter-pixel gap control may lead to a decreased fill factor; (2) the gate lines in an active matrix driving scheme are opened one by one. This scanning update method makes the product of panel resolution and each line's signal time (gate opening + data addressing time) a constant, i.e., the frame time. Keeping the same frame rate, an insufficient charging issue appears if the resolution is increased much; (3) the computational burden increases with the display resolution, which may decrease the source input frame rate.

Hardware-focused solutions have been proposed to address some of the above issues. In 2004, Kanazawa et al. proposed to integrate four LCDs with a pixel-offset method to achieve 8K4K performance [106]. Besides, in foveated displays, an additional panel provides ultrahigh resolution at the central region (FoV ~ 20°) [107]. For the stacked methods, an optical system to accommodate the reflective LCoS without bulkiness needs to be carefully designed. In 2008, JVC demonstrated an 8K4K LCoS projection display, but a smaller pixel pitch is still desired [108].

To detour the hardware issues, people attempt to shift the burden from the spatial domain to the time domain. The field-sequential color (FSC) method is commonly adopted in LCoS projection displays [109,110]. That is how the 4K2K resolution is achieved. FSC display is to blink the RGB backlight in time sequence. In other words, each image frame is divided into three monochrome subframes. In this way, the resolution density could be tripled, but the trade-off is a tripled system refresh rate. For example, a 60-Hz display frame rate with FSC design needs 180 Hz color field rate. Recently, the image shifting and overlaying design were also demonstrated in AR displays [105,111]. By generating two light fields in each frame, the perceived resolution is enhanced. Similar to an FSC display, the required field rate is doubled.

In real-time holographic displays with phase-only LCoS, the required resolution is different from 6K6K as in intensity modulation. Taking phase grating as an example, the FoV correlates to the maximum diffraction angle, which is inversely proportional to the pixel pitch. On the other hand, the angular resolution is inversely proportional to the panel size according to the theory of Fourier transform. In the path light projected to human eyes, the applied optical system adds on factors to the inversely proportional relationships, so that the FoV and angular resolution could be increased or reduced while keeping the same ratio. To obtain a larger FoV with finer angular resolution, AR displays call for higher spatial resolution LCoS panels. Fitting in the same panel size for compact form factor, the pixel pitch decreases as the resolution increases. The state-of-the-art pixel pitch has reached 2.5~3 μm. Ideally, AR displays call for a 1-μm pitch. Challenges of fabricating such a small pitch come from both the electrical and optical sides. The limited space in the pixel area could accommodate very few transistors and storage capacitors, resulting in a low applicable voltage on the LC layer. The fringing field effect and low fill factor impair the throughput optical power. Presently, the relatively large pixel size and limited resolution provided by commercial LCoS panels cannot provide a satisfactory 3D image quality. An upgrade in hardware or new optical system design is highly desirable.

4.2. Augmented Reality Head-Up Displays

Head-up display (HUD) was first developed for military aircraft. Nowadays, it is extended to automobile driving assistance. By projecting a virtual image onto the windshield, the driver does not need to head down to watch for the information displayed on the cockpit cluster. In so doing, the distraction could be minimized. A schematic of today's HUD is shown in Figure 14. By using a double mirror (M1 and M2) reflection and the windshield as the combiner, the displayed information can be integrated with the front scenery. In some cost-effective designs, an additional combiner is employed so that the HUD box works with multiple vehicle models. Current HUD systems are employing transmissive TFT-LCD panels, providing a limited FoV and a relatively short virtual image distance (VID). The typical FoV = 5° × 3° and VID = 2 m still need to be improved keeping the compact package size [112]. Multiple studies suggest a preferred HUD information placement at 5° eccentric to the right and a FoV no larger than 20° × 10° for fast recognition [113,114]. Display registration with the real-world objects is also highly recommended, requiring a variable VID with a wide range [112,113,115,116]. The LCoS-integrated system could realize holographic AR-HUD with a larger FoV (>10° × 5°) and a variable VID (>10 m) [93]. From the LCoS device aspect, the requirements in AR-HUD are somewhat different from that in AR-HMD: (1) the longer VID (>2 m) and the smaller FoV help relieve the burden on resolution; (2) the driving conditions with an intense daylight demand a much higher display luminance for keeping the high ambient contrast ratio [117]; (3) the LCoS device is expected to work in extreme environments with a wide temperature range [59,118]. Among them, the thermal management challenge related to high luminance is a major concern.

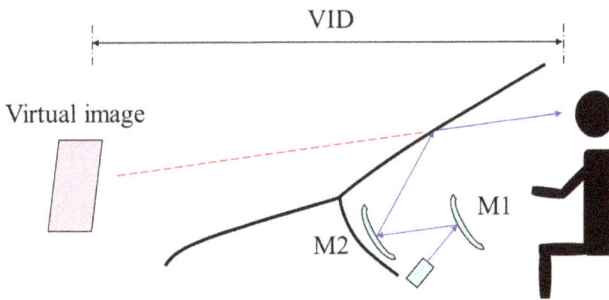

Figure 14. A schematic of a HUD system.

5. Conclusions

After several decades of intensive efforts, LCoS has finally reached several key milestones: (1) 4K2K resolution panels are commercially available and 8K4K prototypes have been demonstrated [108]; (2) presently, a pixel pitch as small as 2.5~3 μm has been developed, while 1 μm is desirable for AR displays; (3) new LC materials with a submillisecond response time for intensity modulation [46,58,59] and ~2 ms response time for phase modulation at 40 °C [46,47] are available, enabling high-frame-rate LCoS with both analog and digital driving; (4) new methods are proposed to suppress the fringing field effect from a large-pixel design to the accommodation of small pixels [78]. Because of the abovementioned achievements and the merits of high fill factor, compact size and light weight, LCoS microdisplay has been integrated into several commercial AR headsets as an intensity modulator [86]. Multi-focal plane [13] and holographic [12] AR-HMDs, and AR-HUDs [93]

Appl. Sci. **2018**, *8*, 2366

with phase-only LCoS are under intensive development. More AR displays incorporating upgraded LCoS panels are expected to emerge in the near distant future.

Author Contributions: General, Y.H. and S.-T.W.; FFE reduction design, E.L.; LC material introduction, R.C.

Funding: This research is funded by Air Force Office for Scientific Research, grant number FA9550-14-1-0279.

Acknowledgments: The authors would like to thank Guanjun Tan, Yun-Han Lee and Tao Zhan for help discussions.

Conflicts of Interest: The authors declare no conflict of interest.

References

1. Margerum, J.D.; Nimoy, J.; Wong, S.Y. Reversible ultraviolet imaging with liquid crystals. *Appl. Phys. Lett.* **1970**, *17*, 51–53. [CrossRef]
2. Beard, T.D.; Bleha, W.P.; Wong, S.Y. Ac Liquid-Crystal Light Valve. *Appl. Phys. Lett.* **1973**, *22*, 90–92. [CrossRef]
3. Ernstoff, M.N.; Leupp, A.M.; Little, M.J.; Peterson, H.T. Liquid crystal pictorial display. *Int. Electron Devices Meet.* **1973**, 548–551. [CrossRef]
4. Efron, U.; Braatz, P.O.; Little, M.J.; Schwartz, R.N.; Grinberg, J. Silicon liquid crystal light valves: Status and issues. *Opt. Eng.* **1983**, *22*, 682–686. [CrossRef]
5. Johnson, K.M.; McKnight, D.J.; Underwood, I. Smart Spatial Light Modulators Using Liquid Crystals on Silicon. *IEEE J. Quantum Electron.* **1993**, *29*, 699–714. [CrossRef]
6. Wu, S.-T.; Wu, C.-S. Mixed-mode twisted nematic liquid crystal cells for reflective displays. *Appl. Phys. Lett.* **1996**, *68*, 1455–1457. [CrossRef]
7. Kuo, C.-L.; Wei, C.-K.; Wu, S.-T.; Wy, C.-S. Reflective direct-view display using a mixed-mode twisted nematic cell. *Jpn. J. Appl. Phys.* **1997**, *36*, 1077–1080. [CrossRef]
8. Alt, P.M. Single crystal silicon for high resolution displays. In Proceedings of the 17th International Display Research Conference, Toronto, ON, Canada, 15–19 September 1997. M19–M28.
9. Sterling, R.D.; Bleha, W.P. D-ILA technology for electronic cinema. *SID Int. Symp. Dig. Tech. Pap.* **2000**, *31*, 310–313. [CrossRef]
10. Melcher, R.L. LCoS-microdisplay technology and applications. *Inf. Disp.* **2000**, *16*, 20–23.
11. Cuypers, D.; de Smet, H.; van Calster, A. VAN LCOS microdisplays: A decade of technological evolution. *J. Disp. Technol.* **2011**, *7*, 127–134. [CrossRef]
12. Maimone, A.; Georgiou, A.; Kollin, J.S. Holographic near-eye displays for virtual and augmented reality. *ACM Trans. Graph.* **2017**, *36*, 85. [CrossRef]
13. Matsuda, N.; Fix, A.; Lanman, D. Focal surface displays. *ACM Trans. Graph.* **2017**, *36*, 86. [CrossRef]
14. Wang, C.; Fu, Q.; Dun, X.; Heidrich, W. Megapixel adaptive optics: Towards Correcting Large-scale Distortions in Computational Cameras. *ACM Trans. Graph.* **2018**, *37*, 115. [CrossRef]
15. Stöhr, J.; Samant, M.G.; Cossy-Favre, A.; Díaz, J.; Momoi, Y.; Odahara, S.; Nagata, T. Microscopic origin of liquid crystal alignment on rubbed polymer surfaces. *Macromolecules* **1998**, *31*, 1942–1946. [CrossRef]
16. Hong, Z.; Zhu, L.; Fu, S.; Tang, M.; Shum, P.; Liu, D. A robust and fast polarimeter based on spatial phase modulation of liquid crystal on silicon (LCoS). In Proceedings of the Asia Communications and Photonics Conference, Hong Kong, China, 19–23 November 2015.
17. Osten, W.; Kohler, C.; Liesener, J. Evaluation and application of spatial light modulators for optical metrology. *Óptica Pura y Aplicada* **2005**, *38*, 71–81.
18. Crossland, W.A.; Wilkinson, T.D.; Manolis, I.G.; Redmond, M.M.; Davey, A.B. Telecommunications applications of LCOS devices. *Mol. Cryst. Liq. Cryst.* **2002**, *375*, 1–13. [CrossRef]
19. Jack, B.; Leach, J.; Ritsch, H.; Barnett, S.M.; Padgett, M.J.; Franke-Arnold, S. Precise quantum tomography of photon pairs with entangled orbital angular momentum. *New J. Phys.* **2009**, *11*, 103024. [CrossRef]
20. Solís-Prosser, M.A.; Arias, A.; Varga, J.J.M.; Rebón, L.; Ledesma, S.; Iemmi, C.; Neves, L. Preparing arbitrary pure states of spatial qudits with a single phase-only spatial light modulator. *Opt. Lett.* **2013**, *38*, 4762–4765. [CrossRef] [PubMed]

21. Sun, P.; Chang, S.; Zhang, S.; Xie, T.; Li, H.; Liu, S.; Wang, C.; Tao, X.; Zheng, Z. Computer-generated holographic near-eye display system based on LCoS phase only modulator. *Proc. SPIE* **2017**, *10396*, 103961J. [CrossRef]

22. Li, Y.-W.; Lin, C.-W.; Chen, K.-Y.; Fan-Chiang, K.-H.; Kuo, H.-C.; Tsai, H.-C. Front-lit LCOS for wearable applications. *SID Int. Symp. Dig. Tech. Pap.* **2014**, *45*, 234–236. [CrossRef]

23. Moon, E.; Kim, M.; Roh, J.; Kim, H.; Hahn, J. Holographic head-mounted display with RGB light emitting diode light source. *Opt. Express* **2014**, *22*, 6526–6534. [CrossRef] [PubMed]

24. Schiekel, M.F.; Fahrenschon, K. Deformation of nematic liquid crystals with vertical orientation in electrical fields. *Appl. Phys. Lett.* **1971**, *19*, 391–393. [CrossRef]

25. Schadt, M.; Helfrich, W. Voltage-dependent optical activity of a twisted nematic liquid crystal. *Appl. Phys. Lett.* **1971**, *18*, 127–128. [CrossRef]

26. Schadt, M. Milestone in the history of field-effect liquid crystal displays and materials. *Jpn. J. Appl. Phys.* **2009**, *48*, 03B001. [CrossRef]

27. Stöhr, J.; Samant, M.G.; Lüning, J.; Callegari, A.C.; Chaudhari, P.; Doyle, J.P.; Lacey, J.A.; Lien, S.A.; Purushothaman, S.; Speidell, J.L. Liquid crystal alignment on carbonaceous surfaces with orientational order. *Science* **2001**, *292*, 2299–2302. [CrossRef] [PubMed]

28. Schadt, M.; Schmitt, K.; Kozinkov, V.; Chigrinov, V. Surface-induced parallel alignment of liquid crystals by linearly polymerized photopolymers. *Jpn. J. Appl. Phys.* **1992**, *31*, 2155–2164. [CrossRef]

29. Schadt, M.; Seiberle, H.; Schuster, A. Optical patterning of multidomain liquid-crystal displays with wide viewing angles. *Lett. Nat.* **1996**, *381*, 212–215. [CrossRef]

30. Gooch, C.H.; Tarry, H.A. The optical properties of twisted nematic liquid crystal structures with twist angles $\leq$90 degrees. *J. Phys. D Appl. Phys.* **1975**, *8*, 1575–1584. [CrossRef]

31. Collings, N.; Davey, T.; Christmas, J.; Chu, D.; Crossland, B. The applications and technology of phase-only liquid crystal on silicon devices. *J. Disp. Technol.* **2011**, *7*, 112–119. [CrossRef]

32. Konforti, N.; Marom, E.; Wu, S.-T. Phase-only modulation with twisted nematic liquid-crystal spatial light modulators. *Opt. Lett.* **1988**, *13*, 251–253. [CrossRef] [PubMed]

33. Wu, S.-T. Design of a liquid crystal based tunable electrooptic filter. *Appl. Opt.* **1989**, *28*, 48–52. [CrossRef] [PubMed]

34. Wu, S.-T.; Wu, C.-S.; Kuo, C.-L. Reflective direct-view and projection displays using twisted-nematic liquid crystal cells. *Jpn. J. Appl. Phys.* **1997**, *36*, 2721–2727. [CrossRef]

35. Wu, S.-T.; Wu, C.-S. A biaxial film-compensated thin homogeneous cell for reflective liquid crystal display. *J. Appl. Phys.* **1998**, *83*, 4096–4100. [CrossRef]

36. Chen, H.; Hu, M.; Peng, F.; Li, J.; An, Z.; Wu, S.-T. Ultra-low viscosity liquid crystal materials. *Opt. Mater. Express* **2015**, *5*, 655–660. [CrossRef]

37. Kirsch, P.; Hahn, A. Liquid crystals based on hypervalent sulfur fluorides: Exploring the steric effects of ortho-fluorine substituents. *Eur. J. Org. Chem.* **2005**, *2005*, 3095–3100. [CrossRef]

38. Chen, R.; Jiang, Y.; Li, J.; An, Z.; Chen, X.; Chen, P. Dielectric and optical anisotropy enhanced by 1,3-dioxolane terminal substitution on tolane-liquid crystals. *J. Mater. Chem. C* **2015**, *3*, 8706–8711. [CrossRef]

39. Chen, R.; Zhao, L.; An, Z.; Chen, X.; Chen, P. Synthesis and properties of allyloxy-based tolane liquid crystals with high negative dielectric anisotropy. *Liq. Cryst.* **2017**, *44*, 2184–2191. [CrossRef]

40. Kirsch, P.; Bremer, M.; Taugerbeck, A.; Wallmichrath, T. Difluorooxymethylene-bridged liquid crystals: A novel synthesis based on the oxidative alkoxydifluorodesulfuration of dithianylium salts. *Angew. Chem. Int. Ed.* **2001**, *40*, 1480–1484. [CrossRef]

41. Yang, X.; Mo, L.; Hu, M.; Li, J.; Li, J.; Chen, R.; An, Z. New isothiocyanato liquid crystals containing thieno[3,2-b]thiophene central core. *Liq. Cryst.* **2017**, *45*, 1294–1302. [CrossRef]

42. Chen, R.; An, Z.; Li, F.; Chen, X.; Chen, P. Synthesis and physical properties of tolane liquid crystals containing 2,3-difluorophenylene and terminated by a tetrahydropyran moiety. *Liq. Cryst.* **2016**, *43*, 564–572. [CrossRef]

43. Lee, S.H.; Bhattacharyya, S.S.; Jin, H.S.; Jeong, K.U. Devices and materials for high-performance mobile liquid crystal displays. *J. Mater. Chem.* **2012**, *22*, 11893–11903. [CrossRef]

44. Li, J.; Peng, Z.; Chen, R.; Li, J.; Hu, M.; Zhang, L.; An, Z. Investigation of terminal olefin in the isothiocyanatotolane liquid crystals with alkoxy end group. *Liq. Cryst.* **2018**, *45*, 1498–1507. [CrossRef]

45. Gauza, S.; Wang, H.; Wen, C.-H.; Wu, S.-T.; Seed, A.J.; Dabrowski, R. High birefringence isothiocyanato tolane liquid crystals. *Jpn. J. Appl. Phys.* **2003**, *42*, 3463–3466. [CrossRef]

46. Huang, Y.; He, Z.; Wu, S.-T. Fast-response liquid crystal phase modulators for augmented reality displays. *Opt. Express* **2017**, *25*, 32757–32766. [CrossRef]

47. Chen, R.; Huang, Y.; Li, J.; Hu, M.; Li, J.; Chen, X.; Chen, P.; Wu, S. High-frame-rate liquid crystal phase modulator for augmented reality displays. *Liq. Cryst.* **2018**. [CrossRef]

48. Chen, Y.; Sun, J.; Xianyu, H.; Wu, S.-T.; Liang, X.; Tang, H. High birefringence fluoro-terphenyls for thin-cell-gap TFT-LCDs. *J. Disp. Technol.* **2011**, *7*, 478–481. [CrossRef]

49. Parri, O.; Wittek, M.; Schroth, D.; Canisius, J. New liquid crystals for light guiding application: From automotive headlights to adaptive indoor lighting. *SID Intl. Symp. Dig. Tech. Pap.* **2017**, *48*, 1157–1159. [CrossRef]

50. Arakawa, Y.; Inui, S.; Tsuji, H. Novel diphenylacetylene-based room-temperature liquid crystalline molecules with alkylthio groups, and investigation of the role for terminal alkyl chains in mesogenic incidence and tendency. *Liq. Cryst.* **2018**, *45*, 811–820. [CrossRef]

51. Chen, R.; An, Z.; Wang, W.; Chen, X.; Chen, P. Improving UV stability of tolane-liquid crystals in photonic applications by the ortho fluorine substitution. *Opt. Mater. Express* **2016**, *6*, 97–105. [CrossRef]

52. Li, J.; Li, J.; Hu, M.; Che, Z.; Mo, L.; Yang, X.; An, Z.; Zhang, L. The effect of locations of triple bond at terphenyl skeleton on the properties of isothiocyanate liquid crystals. *Liq. Cryst.* **2017**, *44*, 1374–1383. [CrossRef]

53. Wen, C.-H.; Gauza, S.; Wu, S.-T. Photostability of liquid crystals and alignment layers. *J. Soc. Inf. Disp.* **2005**, *13*, 805–811. [CrossRef]

54. Jiao, M.; Ge, Z.; Song, Q.; Wu, S.-T. Alignment layer effects on thin liquid crystal cells. *Appl. Phys. Lett.* **2008**, *92*, 061102. [CrossRef]

55. Wu, S.-T.; Efron, U. Optical properties of thin nematic liquid crystal cells. *Appl. Phys. Lett.* **1986**, *48*, 624–626. [CrossRef]

56. Chen, Y.; Peng, F.; Wu, S.-T. Submillisecond-response vertical-aligned liquid crystal for color sequential projection displays. *J. Soc. Inf. Disp.* **2013**, *9*, 78–81. [CrossRef]

57. Chen, Y.; Peng, F.; Yamaguchi, T.; Song, X.; Wu, S.-T. High performance negative dielectric anisotropy liquid crystals for display applications. *Crystals* **2013**, *3*, 483–503. [CrossRef]

58. Chen, H.; Gou, F.; Wu, S.-T. Submillisecond-response nematic liquid crystals for augmented reality displays. *Opt. Mater. Express* **2017**, *7*, 195–201. [CrossRef]

59. Peng, F.; Huang, Y.; Gou, F.; Hu, M.; Li, J.; An, Z.; Wu, S.-T. High performance liquid crystals for vehicle displays. *Opt. Mater. Express* **2016**, *6*, 717–726. [CrossRef]

60. Efron, U. *Spatial Light Modulator Technology: Materials, Devices, and Applications*; Marcel Dekker: New York, NY, USA, 1995.

61. Collings, N.; Pourzand, A.R.; Vladimirov, F.L.; Pletneva, N.I.; Chaika, A.N. Pixelated liquid-crystal light valve for neural network application. *Appl. Opt.* **1999**, *38*, 6184–6189. [CrossRef] [PubMed]

62. Kirzhner, M.G.; Klebanov, M.; Lyubin, V.; Collings, N.; Abdulhalim, I. Liquid crystal high-resolution optically addressed spatial light modulator using a nanodimensional chalcogenide photosensor. *Opt. Lett.* **2014**, *39*, 2048–2051. [CrossRef] [PubMed]

63. Solodar, A.; Kumar, T.A.; Sarusi, G.; Abdulhalim, I. Infrared to visible image up-conversion using optically addressed spatial light modulator utilizing liquid crystal and InGaAs photodiodes. *Appl. Phys. Lett.* **2016**, *108*, 021103. [CrossRef]

64. Shcherbin, K.; Gvozdovskyy, I.; Evans, D.R. Optimization of the liquid crystal light valve for signal beam amplification. *Opt. Mater. Express* **2016**, *6*, 3670–3675. [CrossRef]

65. James, R.; Ferná, F.; Day, S.; Komarčević, M.; William, A. Modelling of high resolution phase spatial light modulators. *Mol. Cryst. Liq. Cryst.* **2004**, *422*, 209–217. [CrossRef]

66. Vanbrabant, P.J.M.; Beeckman, J.; Neyts, K.; James, R.; Fernandez, F.A. Optical analysis of small pixel liquid crystal microdisplays. *J. Disp. Technol.* **2011**, *7*, 156–161. [CrossRef]

67. Cerrolaza, B.; Geday, M.A.; Quintana, X.; Otón, J.M. An optical method for pretilt and profile determination in LCOS VAN displays. *J. Disp. Technol.* **2011**, *7*, 141–150. [CrossRef]

68. Fan-Chiang, K.-H.; Wu, S.-T.; Chen, S.-H. Fringing field effect of the liquid-crystal-on-silicon devices. *Jpn. J. Appl. Phys.* **2002**, *41*, 4577–4585. [CrossRef]

69. Fan-Chiang, K.-H.; Wu, S.-T.; Chen, S.-H. Fringing-field effects on high-resolution liquid crystal microdisplays. *J. Disp. Technol.* **2005**, *1*, 304–313. [CrossRef]

70. Vanbrabant, P.J.M.; Beeckman, J.; Neyts, K.; Willman, E.; Fernandez, F.A. Diffraction and fringing field effects in small pixel liquid crystal devices with homeotropic alignment. *J. Appl. Phys.* **2010**, *108*, 083104. [CrossRef]

71. Ji, Y.; Gandhi, J.; Stefanov, M.E. Stefanov Fringe-field effects in reflective CMOS LCD design optimization. *SID Intl. Symp. Dig. Tech. Pap.* **1999**, *30*, 750–753. [CrossRef]

72. Apter, B.; Efron, U.; Bahat-treidel, E. On the fringing-field effect in liquid-crystal beam-steering devices. *Appl. Opt.* **2004**, *43*, 11–19. [CrossRef] [PubMed]

73. Armitage, D.; Underwood, I.; Wu, S.-T. *Introduction to Microdisplays*; John Wiley & Sons: Chichester, UK, 2006.

74. Chou, W.Y.; Hsu, C.H.; Chang, S.W.; Chiang, H.C.; Ho, T.Y. A novel design to eliminate fringe field effects for liquid crystal on silicon. *Jpn. J. Appl. Phys.* **2002**, *41*, 7386–7390. [CrossRef]

75. Gu, L.; Chen, X.; Jiang, W.; Howley, B.; Chen, R.T. Fringing-field minimization in liquid-crystal-based high-resolution switchable gratings. *Appl. Phys. Lett.* **2005**, *87*, 201106. [CrossRef]

76. Li, Y.-W.; Fan-Chiang, K.-H. Active Matrix Structure and Liquid Crystal Display Panel. U.S. Patent 2015/0002795 A1, 1 January 2015.

77. Liao, C.-H. Spatial Light Modulator Reducing Fringing Field Effect. CN Patent 106716238 A, 24 May 2017.

78. Liao, C.-H. Reducing Fringe Field Effect for Spatial Light Modulator. U.S. Patent 2018/0164643 A1, 14 January 2018.

79. Liao, C.-H. Reducing Fringe Field Effect for Spatial Light Modulator. WO Patent 2018/107517 A1, 21 June 2018.

80. Tseng, M.C.; Fan, F.; Lee, C.Y.; Murauski, A.; Chigrinov, V.; Kwok, H.S. Tunable lens by spatially varying liquid crystal pretilt angles. *J. Appl. Phys.* **2011**, *109*, 083109. [CrossRef]

81. Fan, F.; Srivastava, A.K.; Du, T.; Tseng, M.C.; Chigrinov, V.; Kwok, H.S. Low voltage tunable liquid crystal lens. *Opt. Lett.* **2013**, *38*, 4116–4119. [CrossRef] [PubMed]

82. Otón, E.; Escolano, J.M.; Quintana, X.; Otón, J.M.; Geday, M.A. Aligning lyotropic liquid crystals with silicon oxides. *Liq. Cryst.* **2015**, *42*, 1069–1075. [CrossRef]

83. Sun, J.; Wu, S.-T. Recent advances in polymer network liquid crystal spatial light modulators. *J. Polym. Sci. Part B Polym. Phys.* **2013**, *52*, 183–192. [CrossRef]

84. Sun, J.; Chen, Y.; Wu, S.-T. Submillisecond-response and scattering-free infrared liquid crystal phase modulators. *Opt. Express* **2012**, *20*, 20124–20129. [CrossRef] [PubMed]

85. Kress, B.; Starner, T. A review of head-mounted displays (HMD) technologies and applications for consumer electronics. *Proc. SPIE* **2013**, *8720*, 87200A. [CrossRef]

86. Raffle, H.S.; Wang, C.-J. Heads-Up Display. U.S. Patent 009285877 B2, 15 March 2016.

87. Zhang, Q.; Liu, Z.; Zhang, W.; Yu, F. Polarization recycling method for light-pipe-based optical engine. *Appl. Opt.* **2013**, *52*, 8827–8833. [CrossRef] [PubMed]

88. Howarth, P.A. Potential hazards of viewing 3-D stereoscopic television, cinema and computer games: A review. *Ophthalmic Physiol. Opt.* **2011**, *31*, 111–122. [CrossRef] [PubMed]

89. Hoffman, D.M.; Girshick, A.R.; Akeley, K.; Banks, M.S. Vergence—Accommodation conflicts hinder visual performance and cause visual fatigue. *J. Vis.* **2008**, *8*, 33. [CrossRef] [PubMed]

90. Wann, J.P.; Rushton, S.; Mon-Williams, M. Natural problems for stereoscopic depth perception in virtual environments. *Vision Res.* **1995**, *35*, 2731–2736. [CrossRef]

91. Watt, S.J.; Akeley, K.; Ernst, M.O.; Banks, M.S. Focus cues affect perceived depth. *J. Vis.* **2005**, *5*, 834–862. [CrossRef] [PubMed]

92. Hua, H. Enabling focus cues in head-mounted displays. *Proc. IEEE* **2017**, *105*, 805–824. [CrossRef]

93. Mullins, B.; Greenhalgh, P.; Christmas, J. The holographic future of head up displays. *SID Int. Symp. Dig. Tech. Pap.* **2017**, *48*, 886–889. [CrossRef]

94. Zhang, Z.; You, Z.; Chu, D. Fundamentals of phase-only liquid crystal on silicon (LCOS) devices. *Light Sci. Appl.* **2014**, *3*, e213. [CrossRef]

95. Fienup, J.R. Iterative method applied to image reconstruction and to computer-generated holograms. *Opt. Eng.* **1980**, *19*, 297–305. [CrossRef]

96. Fienup, J.R. Phase retrieval algorithms: A comparison. *Appl. Opt.* **1982**, *21*, 2758–2769. [CrossRef] [PubMed]

97. Georgiou, A.; Christmas, J.; Collings, N.; Moore, J.; Crossland, W.A. Aspects of hologram calculation for video frames. *J. Opt. A Pure Appl. Opt.* **2008**, *10*, 035302. [CrossRef]

98. Matusik, W.; Pfister, H. 3D TV: A scalable system for real-time acquisition, transmission, and autostereoscopic display of dynamic scenes. *ACM Trans. Graph.* **2004**, *23*, 814–824. [CrossRef]
99. Chen, H.; Peng, F.; Hu, M.; Wu, S.T. Flexoelectric effect and human eye perception on the image flickering of a liquid crystal display. *Liq. Cryst.* **2015**, *42*, 1730–1737. [CrossRef]
100. Wang, C.; Hsu, R. Digital modulation on micro display and spatial light modulator. *SID Int. Symp. Dig. Tech. Pap.* **2017**, *48*, 238–241. [CrossRef]
101. Worley, W.S., III; Hudson, E.L.; Weatherford, W.T.; Chow, W.H. System and Method for Using Compound Data Words to Reduce the Data Phase Difference between Adjacent Pixel Electrodes. US Patent 006151011 A, 21 November 2000.
102. García-Márquez, J.; López, V.; González-Vega, A.; Noé, E. Flicker minimization in an LCoS spatial light modulator. *Opt. Express* **2012**, *20*, 8431–8441. [CrossRef] [PubMed]
103. Lazarev, G.; Hermerschmidt, A.; Rozhkov, O.V. LC-based phase-modulating spatial light modulators. In Proceedings of the Imaging and Applied Optics, Seattle, WA, USA, 13–17 July 2014.
104. Martínez, F.J.; Márquez, A.; Gallego, S.; Ortuño, M.; Francés, J.; Beléndez, A.; Pascual, I. Electrical dependencies of optical modulation capabilities in digitally addressed parallel aligned liquid crystal on silicon devices. *Opt. Eng.* **2014**, *53*, 067104. [CrossRef]
105. Lee, Y.-H.; Zhan, T.; Wu, S.-T. Enhancing the resolution of a near-eye display with a Pancharatnam–Berry phase deflector. *Opt. Lett.* **2017**, *42*, 4732–4735. [CrossRef] [PubMed]
106. Kanazawa, M.; Hamada, K.; Kondoh, I.; Okano, F.; Haino, Y.; Sato, M.; Doi, K. An ultrahigh-definition display using the pixel-offset method. *J. Soc. Inf. Disp.* **2004**, *12*, 93–103. [CrossRef]
107. Tan, G.; Lee, Y.-H.; Zhan, T.; Yang, J.; Liu, S.; Zhao, D.; Wu, S.-T. Foveated imaging for near-eye displays. *Opt. Express* **2018**, *26*, 25076–25085. [CrossRef]
108. Sterling, R. JVC D-ILA high resolution, high contrast projectors and applications. In Proceedings of the IPT/EDT '08, Los Angeles, CA, USA, 9–10 August 2008; ACM: New York, NY, USA, 2008.
109. Hasebe, H.; Kobayashi, S. A full-color field sequential LCD using modulated backlight. *SID Int. Symp. Dig. Tech. Pap.* **1985**, *16*, 81–83.
110. Huang, Y.P.; Lin, F.C.; Shieh, H.P.D. Eco-displays: The color LCD's without color filters and polarizers. *J. Disp. Technol.* **2011**, *7*, 630–632. [CrossRef]
111. Zhan, T.; Lee, Y.-H.; Wu, S.-T. High-resolution additive light field near-eye display by switchable Pancharatnam–Berry phase lenses. *Opt. Express* **2018**, *26*, 4863–4872. [CrossRef] [PubMed]
112. Gabbard, J.L.; Fitch, G.M.; Kim, H. Behind the glass: Driver challenges and opportunities for AR automotive applications. *Proc. IEEE* **2014**, *102*, 124–136. [CrossRef]
113. Haeuslschmid, R.; Shou, Y.; O'Donovan, J.; Burnett, G.; Butz, A. First steps towards a view management concept for large-sized head-up displays with continuous depth. In Proceedings of the Automotice'UI 16, Ann Arbor, MI, USA, 24–26 October 2016; ACM: New York, NY, USA, 2016; pp. 1–8.
114. Yoo, H.; Tsimhoni, O.; Watanabe, H.; Green, P.; Shah, R. *Display of HUD Warnings to Drivers: Determining an Optimal Location*; Technical Reports; UMTRI-99-9; University of Michiagan Transporation Research Institute: Ann Arbor, MI, USA, 1999.
115. Plavšic, M.; Duschl, M.; Tönnis, M.; Bubb, H.; Klinker, G. Ergonomic design and evaluation of augmented reality based cautionary warnings for driving assistance in urban environments. In Proceedings of the 17th World Congress on Ergonomics, Beijing, China, 9–14 August 2009; Chinese Ergonomics Society: Beijing, China, 2009.
116. Sato, A.; Kitahara, I.; Kameda, Y.; Ohta, Y. Visual navigation system on windshield head-up display. In Proceedings of the 13th ITS World Congress, London, UK, 8–12 October 2006.
117. Chen, H.; Tan, G.; Wu, S.-T. Ambient contrast ratio of LCDs and OLED displays. *Opt. Express* **2017**, *25*, 33643–33656. [CrossRef]
118. Peng, F.; Gou, F.; Chen, H.; Huang, Y.; Wu, S.-T. A submillisecond-response liquid crystal for color sequential projection displays. *J. Soc. Inf. Disp.* **2016**, *24*, 241–245. [CrossRef]

applied
sciences

MDPI

Review

Displays Based on Dynamic Phase-Only Holography

Jamieson Christmas and Neil Collings *

Envisics Ltd., Milton Keynes MK5 8PG, UK; Jamieson.Christmas@envisics.com
* Correspondence: neil.collings@envisics.com; Tel.: +44-1908-886-220

Received: 15 March 2018; Accepted: 24 April 2018; Published: 27 April 2018

Abstract: Static holographic displays of high quality are works of art. We are creating commercial displays using dynamic phase-only spatial light modulators (SLMs). The main advantages of this approach are light efficiency and fault tolerance. When polarized lasers are used as the illumination source, there is no requirement for polarizers in the light engine. Moreover, the illumination beam can be directed towards bright points of the image and away from dark regions. Due to the many-to-one correspondence between the pixels in the SLM and the points in the image, faults in high complexity SLMs will be annealed in the image. Compared with normal displays where etendue is of overriding importance for light efficiency, holographic displays favor small pixel devices. Smaller pixel devices generate a larger reconstruction which improves the etendue for a second stage imaging system.

Keywords: holography; kinoform; head-up display; spatial light modulator; computer generated hologram

1. Introduction

Holographic displays are one of the strong contenders for realizing true 3D displays. The first display holograms were based on off-axis holograms recorded on a photographic plate [1]. In order to develop dynamic displays, computer-generated holograms (CGHs) are written on to a spatial light modulator (SLM). The current complexity of SLMs limits the size of the CGH. However, prototypes have been developed using SeeReal [2], Microsoft Research [3], Light Blue Optics [4], and the European Real3D project [5]. This work has focused on the reconstruction of 2D scenes and has grown from student work [6] to commercialization within 10 years.

2. Fourier and Fresnel Holography

The first display holograms were off-axis Fresnel holograms, which captured interference between an object wave and a plane wave reference beam. Since the hologram is the viewing window, there was an interest in using holographic recording media of a large size (see Figure 1) [7].

The 0 order beam in Figure 1 has the same size as the illumination beam. Therefore, a high resolution holographic recording was important so that the angular separation of the orders was sufficient. Unfortunately, SLMs are not available with the same size and resolution as the photographic plates used in Reference [1].

In order to capitalize on the limited SBWP available from the SLM, a Fourier arrangement is used for the hologram replay. A Fourier hologram is the hologram that is recorded when the plane reference beam is focused on a spot at the same distance as the object. The beams interfering at the hologram plane will then have approximately equal radii of curvature. This reduces the interference fringing between the two beams, which demands a share of the SBWP. When illuminated by a plane wave, the Fourier hologram will reconstruct the object at infinity. If the Fourier hologram is reconstructed with a plane wave and a lens is placed in the reconstructed beam, the reconstruction is placed in the focal plane of the lens (see Figure 2).

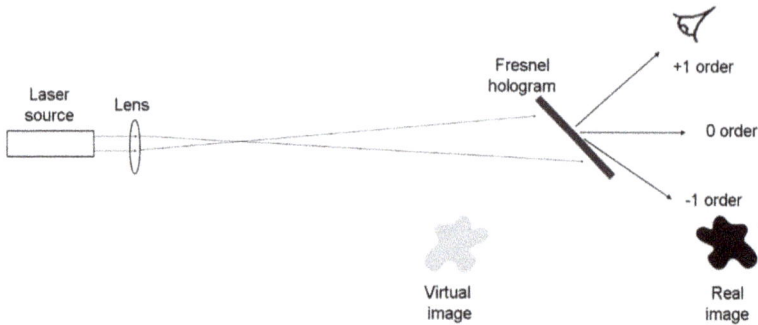

Figure 1. Viewing geometry for an off-axis Fresnel hologram.

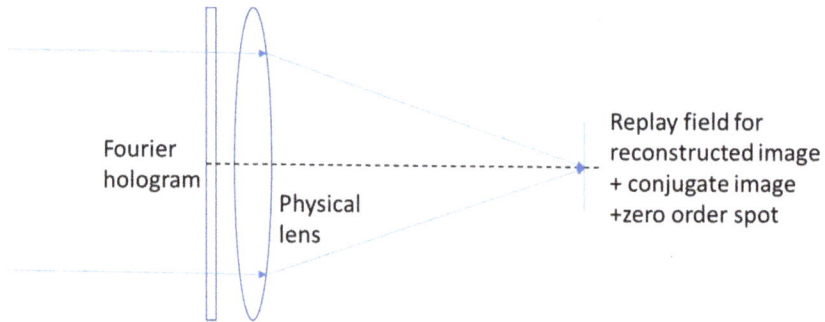

Figure 2. Fourier hologram with a physical lens.

The advantage of the Fourier arrangement is that the image size can be adjusted by varying the focal length of the physical lens. Unfortunately, increasing the image size reduces the field of view (FOV) [8] so that additional components are required to increase the field of view (see Section 5). The zero order, which is formed by all the light transmitted by the hologram that is not diffracted, is focused to a spot at the focal point of the lens. In addition, the many-to-one correspondence, which promotes the fault tolerance, is accentuated in the Fourier case compared with the Fresnel hologram.

In order to separate the reconstructed image from the conjugate image in the Fourier arrangement, a software lens is superimposed on the hologram [9]. The nature of the software lens depends on the modulation used in the hologram. For a binary amplitude modulation, the lens is a Fresnel zone plate. For a binary phase hologram, it is a phase zone plate and, for an analogue phase hologram, it is a phase Fresnel lens. In Figure 3, the positions of the reconstructed image and the conjugate image are shown for the case of a weak diverging software lens superimposed on the Fourier hologram. The reconstructed image is separated from both the zero order spot and the conjugate image. The zero order spot can be readily removed by spatial filtering. The size of the reconstructed image is proportional to the distance of the replay field from the hologram. The Fourier arrangement was used by Qinetiq in their Active Tiling 3D display system [8].

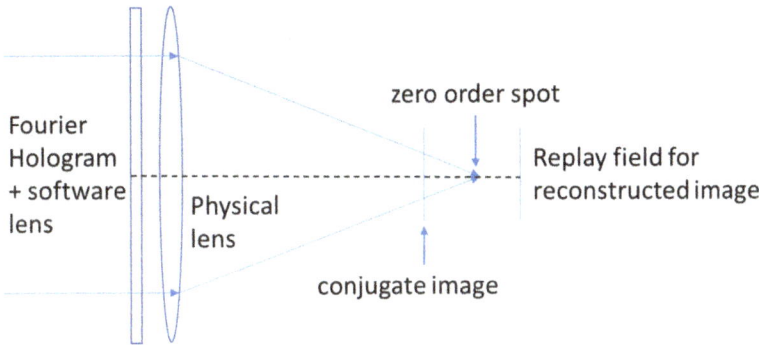

Figure 3. Fourier hologram with a physical lens and software lens.

3. Spatial Light Modulators (SLMs)

The SLM is responsible for the shape of the wave front in the display system which, by free-space propagation, determines the resultant display. In order to achieve a fully complex wave front, which varies in amplitude and phase at each point on the wave front, it is currently necessary to cascade or superimpose two SLMs using beam splitters or a grating or combine two areas from a single SLM. The value of complex coding would be minimal computation and near perfect reconstruction. The sub-systems required for complex coding are currently being researched in a number of laboratories [2,10]. It has been possible to achieve initial holographic display systems at the full resolution of the SLM by using limited modulation devices in particular phase-only electrically addressed devices [11] and amplitude-only optically addressed [9] devices. The former are principally Liquid Crystal on Silicon (LCOS) SLMs and the latter are Liquid Crystal Light Valves (LCLVs) [12]. Due to the technical maturity of the LCOS SLMs, they are the focus of the present paper.

LCOS devices are composed of a liquid crystal layer above a semiconductor backplane. The front substrate is a cover glass coated with a transparent conductor to act as a common electrical conductor. The lower substrate is a CMOS backplane with an additional outer aluminum layer that is used to create a pixelated mirror surface. These mirrors also act as the pixel electrodes. The LCOS devices have the advantage that the signal lines, the gate lines, and the transistors are below the mirrored surface, which results in a high fill factor (typically greater than 90%). Relatively conservative 0.18 and 0.35 micron processes with 5 V drive capability and high yield are used by the LCOS fabrication houses. Additionally, the small feature sizes enabled by CMOS processing allow for very small pixel sizes. The maximum diffraction angle of the spatial light modulator is inversely proportional to the pixel pitch.

The perfect phase modulating spatial light modulator (PSLM) would ideally have characteristics similar to a holographic film including a continuous modulation surface (i.e., no pixels) and analogue phase modulation that is capable of greater than 2π phase delay in a large number of steps. Phase-only holograms have been created using liquid crystal TV screens [13]. They have been developed over the last decade so that there is now a reasonably sized supply [11]. All current multilevel phase-only modulating devices use 'Electrically Controlled Birefringence' (ECB), which is shown in Figure 4.

ECB was the first electro-optic effect to be observed in liquid crystals [14]. It is the birefringence reduction in a planar aligned nematic liquid crystal (NLC) layer upon application of an electric field. The birefringence reduction allows the phase modulation of the reflected beam to be reduced by 2π when the polarization of the incident light is parallel to the alignment direction. The birefringence reduction is due to a tilted director where the angle of tilt is determined by two competing torques including the elastic torque within the material that tries to align the director parallel to the alignment direction and a field-induced torque that tends to align the director parallel to the field. When both torques exist, there is a competition between them and the director aligns to minimize the total energy.

At the edge of the pixel, the electric field has a component parallel to the cell substrate surfaces, which is known as the lateral field, the non-uniform field [15], or the inhomogeneous field [16]. It is now more commonly referred to as a fringing field. In addition, there is a reverse tilt declination over the driven pixel due to the conflict between the pre-tilt and the director rotation because of the vertical electric field.

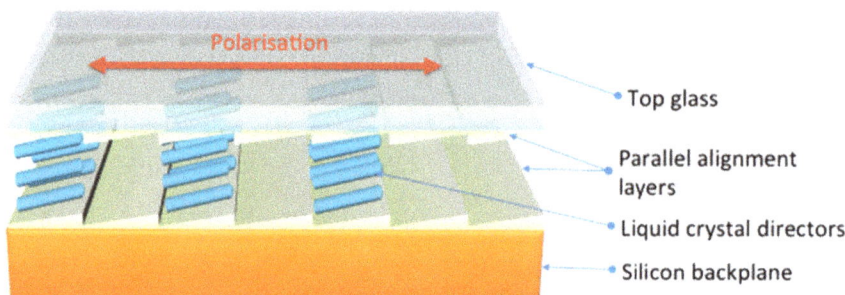

Figure 4. Electrically Controlled Birefringence Mode.

There are two methods for matrix addressing the CMOS pixel array. The analogue address known as the Dynamic Random Access Memory (DRAM) is used for driving the nematic LC to a number of grey levels. The digital address known as Static Random Access Memory (SRAM) applies digital bit array sequences to the LCOS. In the SRAM device, analogue modulation depth is achieved by temporal multiplexing. If the response time of the LC is slow relative to the period of the sequence, then the LC gives an average response, which is the required analogue RMS value. For example, a thick LC layer will provide the required average and give higher phase modulation depth than 2π. However, additional problems arise in the effective resolution of the device and the frame speed.

Due to the growth of the pico-projector and near-eye markets, there has been a push towards high resolution LCOS micro-displays. Developing a phase-only spatial light modulator based upon these existing micro-display backplanes enables the incumbent liquid crystal assembly facilities to be used without significant upfront capital investment.

In small pixel LCOS devices, the NLC layer must be designed to minimize the fringing field. Otherwise, when high resolution patterns are used, the available phase retardation from a pixel falls short of what is intended. The effect on the performance of the device is that the resolution is severely reduced at high spatial frequencies. The design of the NLC layer to minimize the effect of the fringing field is considerably assisted by using a thin NLC layer [16].

The frame speed also benefits from a thin NLC layer. A doubling of the cell thickness causes a four times increase in the liquid crystal rise and fall times [17]. For phase only modulation, it is important to delay the illumination of the SLM until the director of the liquid crystal is stable. Therefore, a slower liquid crystal response time directly impacts the available illumination time and reduces the display brightness. The reduced liquid crystal response times may be achieved either by using high birefringence liquid crystals and, thereby, enabling the cell gap to be reduced or by operating the spatial light modulator at more than 45 °C, which lowers the viscosity. The latter option is not available when SRAM backplane devices are used because the reduced viscosity gives an enhanced phase ripple and consequent noise in the replay field.

In practice, all of these techniques are utilized at least partially to achieve an acceptable level of performance from the PSLM. The finished device is equipped with a ceramic backplane for low thermal resistance (see Figure 5). The limited space bandwidth product (SBWP) of current SLMs (1–10 megapixels) does not allow overly complex, off-axis holograms. Therefore, current implementations are restricted to 2D displays.

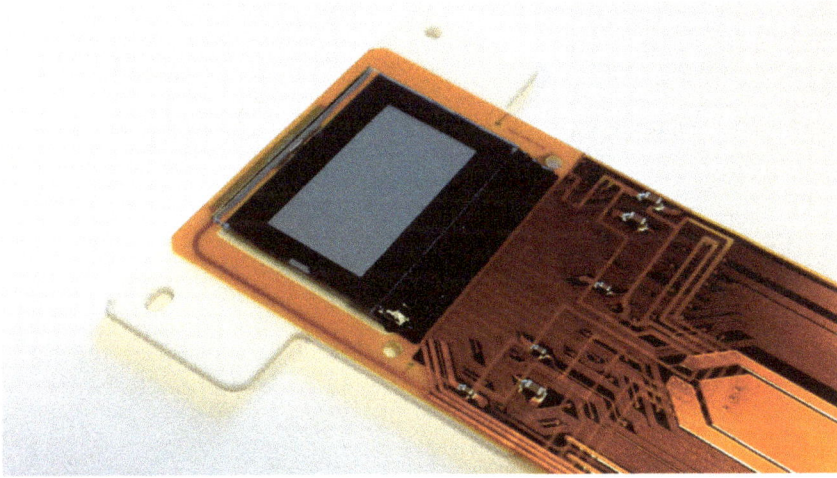

Figure 5. High performance, phase-only SLM.

4. Algorithms

Computational holography makes high demands on computational resources. Therefore, a large effort has been deployed in fine-tuning the algorithms in order to minimize the computational burden. A large reduction of computational efforts has been achieved by incorporating into the design, the limited resolution of the human visual system (HVS), and the position of the viewer (defined eye box or eye tracking device). The extra computational resources can be deployed for a high contrast, low noise image, or a display, which must be updated at video rates or faster. The following discussion tracks the advances, which have been made in reducing the computational burden and then summarizes our efforts with the production of holograms for 2D display in consumer products.

The Active Tiling 3D display system of Qinetiq [9] employed a variant of the diffraction-specific (DS) algorithm [18]. The DS algorithm advanced computational holography by providing solutions in four problem areas including noise, speed, 3D object definition, and encoding. The noise problem was the appearance of the DC term in the hologram replay, which was solved by using an off-axis reference beam in Reference [1]. This solution places significant demands on the limited SBWP of SLMs and is currently not practical for dynamic holography. These DC terms are omitted in the DS algorithm. The speed of the computation was increased by a factor of over 100 by taking account of the limited resolution of the HVS. The object in Reference [1] existed in space while computer models exist as point-clouds or patch-models. Algorithms, which sum the contributions from the object elements (points or patches) in these computer models over the whole hologram, are computationally expensive. The DS algorithm improved the speed of computation by pre-computing the basis fringes for the diffraction of light in discrete directions so that the run-time computation is reduced to table look-up and multiplication by the desired replay field amplitudes. Crucially, amplitude modulated fringes allow a linear superposition while if the basis fringes were phase coded, then linear superposition is only valid for small phase modulation depth (Appendix B in ref. [18]). Amplitude modulation of the 3D holographic display has been preferred to this day [19].

The wave front-recording plane (WRP) method [19] is based on adding the contributions from the object elements over small areas on a plane close to the object. A look-up table can be adopted for increased speed in the first step. The Fresnel diffraction from the WRP to the hologram plane is calculated using the fully complex FFT algorithm so that the modulation format of the computed hologram can be chosen as amplitude, phase, or fully complex. However, the computational effort required for 3D objects still places a demonstrator as a lab project rather than a consumer item. The approach taken here is to begin the work with 2D displays in consumer products and extends the work towards 3D when consumer acceptance has improved to the extent that design of a custom ASIC can be commercially viable.

The initial holographic display systems are based on phase-only holograms (or kinoforms). A phase-only hologram, which is computer-generated was announced in 1969 [20], and an iterative algorithm for the design of these kinoforms known as the GS algorithm was presented a few years later [21]. The advantage of the phase-only hologram for display is that a significant portion of the light energy arrives at the image plane. This enables a more efficient display system but comes at the expense of a significant noise floor. The GS iterative phase retrieval algorithm for calculating the kinoform produces relatively noisy images with poor contrast. This is also known as the Iterative Fourier Transform Algorithm (IFTA) [22]. We have developed a new generation of iterative phase retrieval algorithms that deliver outstanding image quality, which was exemplified in Reference [23]. The latest generation of iterative algorithms can now achieve satisfactory display quality within 10 iterative cycles including a mean square error of less than 1 and a contrast ratio of greater than 5000:1 [24].

5. System Application

The two methods of projection presented in Figure 6 are fundamentally different in terms of the device, optics, and projection image quality. Conventional displays employ imaging where a large size of micro-display is preferable in order to increase the etendue of the system. The NA of the optics adjacent to the micro-display is large for the same reason. In diffractive projection, the light throughput is, in principle, not limited by etendue considerations. Diffractive spread from the SLM to the screen creates the projected image. The complex projection optics of the imaging system is reduced to a Fourier lens. Since large diffraction angles allow the volume of the entire display engine to be reduced, a small pixel size is highly desirable. The quality of the projected image in an imaging system can be markedly influenced by pixel defects and inter-pixel gaps. In contrast, the projected image in a diffractive projection is fault tolerant with regard to the SLM and the inter-pixel gaps do not influence the image. Since coherent light is used in the diffractive system, there is a degree of speckle in the image.

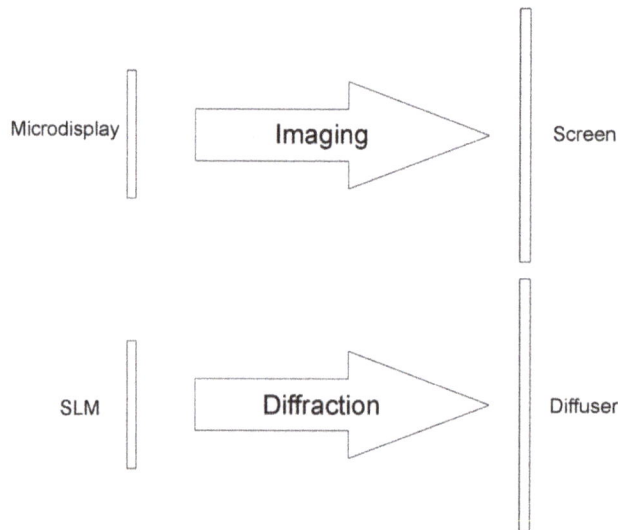

Figure 6. High performance, phase-only SLM.

The product of the image size and the field of view (FOV) in a Fourier transform system is given by the product of the number of pixels in the SLM and the wavelength of the coherent light illumination [8]. For 3D direct view, two distinct approaches have emerged. The first is to increase the number of pixels in the SLM by using Active Tiling from a fast electrically addressed SLM onto an optically addressed SLM [8]. The second is to settle for the number of pixels in current SLMs to restrict the view presented to the eye using eye tracking and increased optics associated with the display [25]. For 2D display, based on the SLMs discussed in Section 3, the product of the image size and the field of view is still far smaller than the field of view that is commercially acceptable. Therefore, a two-stage design is adopted where the first stage results in the projected image on a diffuser and the second stage is an imaging system. The FOV of the diffuser is engineered to give the correct aspect ratio for the FOV of the final image. The image size is magnified in the imaging system so that the FOV*image size product is an invariant. The brightness of the magnified image is consequently equal to the brightness at the diffuser if there is no attenuation in the imaging system. This is the brightness given in our previous study [23]. In order to reduce the speckle in the final projected image, the diffuser is rotated.

The design of the diffraction system is relatively simple in concept. A coherent light source is collimated and illuminates an LCOS device, which displays the phase-only hologram and the reflected beam is brought to a focus by a lens. No polarizers are required if the light source is polarized, e.g., a diode laser. Laser-based light sources offer an improved color gamut in comparison with white LED light sources. In order to provide a display that is competitive in the market place, a large effort is made in the system detail. This will be outlined in this section.

The beam-shaping optics for the diode laser should be designed for high efficiency with minimum wave front error at the LCOS. It is often convenient to tilt the LCOS plane with respect to the collimated beam in order to avoid the use of either a beam splitter or an X-cube. The former attenuates the throughput and the latter is costly. In a conventional amplitude-modulating display, this would lead to a tilted image plane and additional complications for the optics downstream of this plane. The many-to-one correspondence again allows flexibility in the design of the hologram so that the image plane can be perpendicular to the optical axis. The major benefit of a phase only holographic head up display in comparison with an LED/TFT solution for a typical HUD specification is in the throughput and this was detailed in Reference [23].

Due to the many-to-one correspondence between the pixels in the SLM and the points in the image, there is no longer a direct relationship between the pixels on the LCOS device and the pixels in the replay field. The size of the pixels in the replay field is determined by the physical process of optical diffraction. The aperture of the LCOS device is the limiting aperture so that the replay field pixels are ellipses with their long axis vertical, which corresponds to the short axis of the LCOS. When the illumination of the LCOS is uniform, these spots are sinc squared functions. This results in the most compact spot. Uniform illumination with no concomitant light loss is an "ideal" objective, which is difficult to achieve in practice. In order to assess the approaches to beam shaping, Romero and Dickey introduced a dimensionless parameter β [26].

$$\beta = \frac{2\pi RD}{\lambda f}$$

where R is the dimension of the input beam, D is the dimension of the output beam, λ is the wavelength, and f is the distance between input and output, which is commonly the focal length of the collimating lens. The $1/e^2$ radius of the Gaussian waist is a convenient measure for R and the half width of the LCOS aperture is convenient for D. The small $1/e^2$ radius of a laser diode favors large radius output beams. The relatively small aperture of the LCOS leads to a small β. The design of the beam shaping optics becomes more difficult as β decreases. The collimation optic design minimizes both the loss of intensity and the wave front error at the LCOS. The latter minimizes spot size increases due to pupil aberrations.

A good compromise is to expand the Gaussian beam profile and apodise it using the LCOS aperture. The illumination beam profile is an elliptical Gaussian beam with a $1/e^2$ waist radius, w_1, at the corner of the LCOS. The horizontal and vertical waists, w_{1x} and w_{1y}, are in the same ratio as the rectangular aperture of the LCOS (approx. 2:1) and the perimeter of the active area is 0.707 w_{1x} and 0.707 w_{1y}. The spot size (3 dB diameter) will then be seen in the equation below [27].

$$\text{Spot size}_x = \frac{1.4\lambda \times f}{1.414 w_{1x}}$$

$$\text{Spot size}_y = \frac{1.4\lambda \times f}{1.414 w_{1y}}$$

The power captured by the LCOS aperture at this apodisation is 71% of the incident beam. One of the differentiating features of a phase-only holographic display is its ability to redirect light in the formation of the target image. For low information content images such as those used in head up displays, this causes a concentration of the light energy in the utilized pixels, which significantly increases the brightness of the illuminated areas of the display. This is termed "Holographic Gain." Taking the reciprocal of the pixel utilization approximates the Holographic Gain. A typical figure for the Holographic Gain in a head up display application is 5.

The size and resolution of a device such as an SLM can be expressed in a single parameter, which is a dimensionless product of the two and the Space Bandwidth Product (SBWP). The size of the LCOS device shown in Figure 7 is LΔuxMΔv. The resolution in the x-direction is $(\Delta u)^{-1}$ and the resolution in the y-direction is $(\Delta v)^{-1}$. Therefore, the SBWP is equal to the number of pixels in the device, LxM, when the resolution is not reduced by, for example, fringing fields. The spot separation in the replay field can be controlled by hologram repetition [28], which is also illustrated in Figure 7. If the size of the hologram is C $\times$ B pixels within an LCOS aperture of M $\times$ L pixels, then the repetition ratios are M/C and L/B in the x-axes and y-axes. We employ a repetition ratio of around 1.4 to give a pleasing replay field display. Since the separation of the spots is (λf/BΔu, λf/CΔv), the maximum resolution in the replay field in an ideal case will be (N_x/1.4, N_y/1.4). The ideal case assumes a perfect Gaussian beam, no speckle noise, and a perfect optical system.

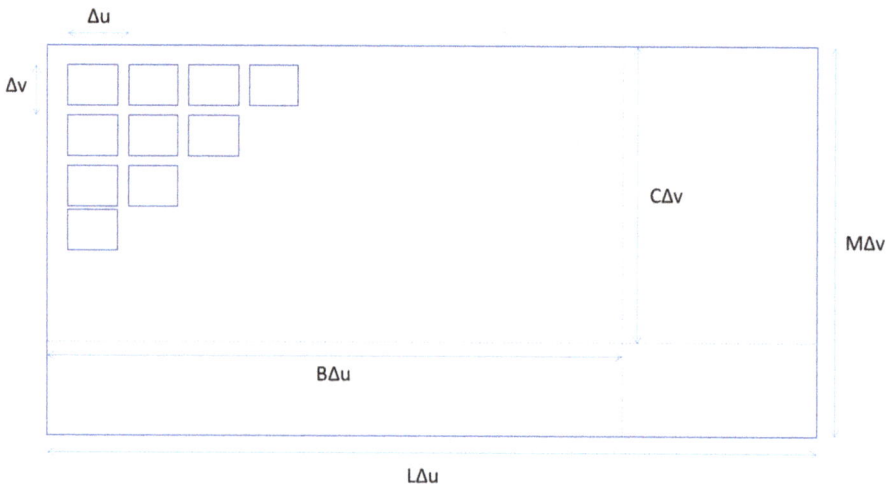

Figure 7. Pixel layout of LCOS.

Appl. Sci. **2018**, *8*, 685

In addition to improving the image quality, the size of the hologram has been reduced, which facilitates the speed of computation.

Defining the spot separation is important for reducing the overlap between the side lobes of the spots, which produce unnecessary speckle noise. It will also define the maximum resolution of the display, which is the image size divided by the spot separation ($N_x/1.4$, $N_y/1.4$). The actual resolution is determined principally by the performance of the LCOS device. As discussed previously, we use thin NLC layers in order to guarantee a phase modulation amplitude between neighboring pixels of the device, which matches the intended pixels by the algorithm used to compute the hologram. Therefore, pixels at the extremity of the image field benefit from the full modulation capability of the backplane.

6. Head-Up Display

The first commercial application of this technology has been the holographic head-up display (HUD) [23]. The holographic system is capable of delivering identical levels of performance at much lower power levels or significantly brighter images than those currently possible with LED backlights [23] and those that use LED front illumination. Advantageously, the holographic display allows defects in the display device to be annealed, software control of the pixellation of the display, and holographic gain. The LCOS SLMs allow the largest format of all 2D display technologies for the required power levels of the HUD. Over the past 10 years, patents in this area of display technology have increased sixteen folds.

7. Conclusions

Phase-only holography is a compelling technology for both 2D and 3D displays. 2D head-up displays are a relatively low information content display with particular requirements, which can be fulfilled by current phase-only display devices. Initial results on 3D objects are promising [25,29,30].

Author Contributions: J.C. developed the first commercial application of the holographic head-up display technology around which this paper is structured; N.C. wrote the paper.

Conflicts of Interest: The authors declare no conflict of interest.

References

1. Leith, E.N.; Upatnieks, J. Wavefront Reconstruction with Diffused Illumination and Three-Dimensional Objects. *J. Opt. Soc. Am.* **1964**, *54*, 1295–1301. [CrossRef]
2. Häussler, R.; Gritsai, Y.; Zschau, E.; Missbach, R.; Sahm, H.; Stock, M.; Stolle, H. Large real-time holographic 3D displays: Enabling components and results. *Appl. Opt.* **2017**, *56*, F45–F52. [CrossRef] [PubMed]
3. Maimone, A.; Georgiou, A.; Kollin, J. Holographic near-eye displays for virtual and augmented reality. *ACM Trans. Gr.* **2017**, *36*. [CrossRef]
4. Buckley, E. 70.2: Invited Paper: Holographic Laser Projection Technology. *SID Symp. Dig. Tech. Pap.* **2008**, *39*, 1074–1079. [CrossRef]
5. Finke, G.; Kozacki, T.; Kujawińska, M. Wide viewing angle holographic display with a multi-spatial light modulator array. *Proc. SPIE* **2010**, 77230A. [CrossRef]
6. Christmas, J.L. Real Time Holography for Displays. Ph.D. Thesis, University of Cambridge, Cambridge, UK, 2009.
7. Collier, R.J.; Burckhardt, C.B.; Lin, L.H. *Optical Holography*; Academic Press/Bell Telephone Laboratories: Murray Hill, NJ, USA, 1971.
8. Slinger, C.; Cameron, C.; Stanley, M. Computer-Generated Holography as a Generic Display Technology. *Computer* **2005**, *38*, 46–53. [CrossRef]
9. Moreno, I.; Campos, J.; Gorecki, C.; Yzuel, M.J. Effects of Amplitude and Phase Mismatching Errors in the Generation of a Kinoform for Pattern Recognition. *Jpn. J. Appl. Phys.* **1995**, *34*, 6423. [CrossRef]
10. Lee, H.-S.; An, J.; Seo, W.; Choi, C.-S.; Kim, Y.-T.; Sung, G.; Seo, J.; Song, H.; Kim, S.; Kim, H.; et al. 54-2: Invited Paper: Holographic Display and its Applications. *SID Symp. Dig. Tech. Pap.* **2017**, *48*, 808–810. [CrossRef]

11. Collings, N.; Christmas, J.L.; Masiyano, D.; Crossland, W.A. Real-Time Phase-Only Spatial Light Modulators for 2D Holographic Display. *J. Disp. Technol.* **2015**, *11*, 278–284. [CrossRef]
12. Collings, N. Optically addressed spatial light modulator for 3D Display. *J. Nonlinear Opt. Phys. Mater.* **2011**, *20*, 453–457. [CrossRef]
13. Pourzand, A.; Favre, S.; Collings, N. *The Optimization of a LCTV for Phase-Only Filtering*; EOS Topical Meetings Digests Series Volume 6; Optics and Information Paper 3.5; EOS Topical Meetings: Teddington, UK, 1995.
14. Freedericksz, V.; Zolina, V. Forces causing the orientation of an anisotropic liquid. *Trans. Faraday Soc.* **1933**, *29*, 919–930. [CrossRef]
15. Blinov, L.M.; Chigrinov, V.G. Modulated and nonuniform structures in nematic liquid crystals. In *Electrooptic Effects in Liquid Crystal Materials*; Partially Ordered Systems; Springer: New York, NY, USA, 1994.
16. Chigrinov, G.; Kompanets, I.N.; Vasiliev, A.A. Behaviour of Nematic Liquid Crystals in Inhomogeneous Electric Fields. *Mol. Cryst. Liq. Cryst.* **1979**, *55*, 193–208. [CrossRef]
17. Wu, S.; Efron, U. Electro-Optic Behavior of Thin Nematic Liquid Crystal Cells. *Proc. SPIE* **1986**, *613*, 172–177.
18. Lucente, M. Diffraction-Specific Fringe Computation for Electro-Holography. Ph.D. Thesis, Massachusetts Institute of Technology, Cambridge, MA, USA, 1994.
19. Shimobaba, T.; Nakayama, H.; Masuda, N.; Ito, T. Rapid calculation algorithm of Fresnel computer-generated-hologram using look-up table and wavefront-recording plane methods for three-dimensional display. *Opt. Express* **2010**, *18*, 19504–19509. [CrossRef] [PubMed]
20. Lesem, L.B.; Hirsch, P.M.; Jordan, J.A. The Kinoform: A New Wavefront Reconstruction Device. *IBM J. Res. Dev.* **1969**, *13*, 150–155. [CrossRef]
21. Gerchberg, R.W.; Saxton, W.O. A Practical Algorithm for the Determination of Phase from Image and Diffraction Plane Pictures. *Optik* **1972**, *35*, 237–246.
22. Wyrowski, F.; Bryngdahl, O. Iterative Fourier-transform algorithm applied to computer holography. *J. Opt. Soc. Am. A* **1988**, *5*, 1058–1065. [CrossRef]
23. Christmas, J.; Collings, N. 75-2: Invited Paper: Realizing Automotive Holographic Head Up Displays. *SID Symp. Dig. Tech. Pap.* **2016**, *47*, 1017–1020. [CrossRef]
24. Mullins, B.; Greenhalgh, P.; Christmas, J. 59-5: Invited Paper: The Holographic Future of Head up Displays. *SID Symp. Dig. Tech. Pap.* **2017**, *48*, 886–889. [CrossRef]
25. Leister, N.; Schwerdtner, A.; Fütterer, G.; Buschbeck, S.; Olaya, J.; Flon, S. Full-color interactive holographic projection system for large 3D scene reconstruction. *Proc. SPIE* **2008**, 69110V. [CrossRef]
26. Romero, A.; Dickey, F.M. Mathematical and physical theory of lossless beam shaping. In *Laser Beam Shaping Theory and Techniques*; CRC Press: Boca Raton, FL, USA, 2000; pp. 21–162.
27. Marom, E.; Chen, B.; Ramer, O.G. Spot size of focused truncated gaussian beams. *Opt. Eng.* **1979**, *18*, 79–81.
28. Wyrowski, F.; Hauck, R.; Bryngdahl, O. Computer-generated holography: hologram repetition and phase manipulations. *J. Opt. Soc. Am. A* **1987**, *4*, 694–698. [CrossRef]
29. Zhang, H.; Xie, J.; Liu, J.; Wang, Y. Optical reconstruction of 3D images by use of pure-phase computer-generated holograms. *Chin. Opt. Lett.* **2009**, *7*, 1101–1103. [CrossRef]
30. Bruckheimer, E.; Rotschild, C.; Dagan, T.; Amir, G.; Kaufman, A.; Gelman, S.; Birk, E. Computer-generated real-time digital holography: First time use in clinical medical imaging. *Eur. Heart J. Cardiovasc. Imaging* **2016**, *17*, 845–849. [CrossRef] [PubMed]

applied sciences

MDPI

Article

Programmable Zoom Lens System with Two Spatial Light Modulators: Limits Imposed by the Spatial Resolution

Jeffrey A. Davis [1], Trevor I. Hall [1], Ignacio Moreno [2,*], Jason P. Sorger [1] and Don M. Cottrell [1]

[1] Department of Physics, San Diego State University, San Diego, CA 92182-1233, USA; jeffrey.davis@sdsu.edu (J.A.D.); trevor-hall@hotmail.com (T.I.H.); jpsorger@gmail.com (J.P.S.); thorn@underweedsmanor.net (D.M.C.)

[2] Departamento de Ciencia de Materiales, Óptica y Tecnología Electrónica, Universidad Miguel Hernández de Elche, 03202 Elche, Spain

* Correspondence: i.moreno@umh.es

Received: 10 May 2018; Accepted: 14 June 2018; Published: 20 June 2018

Featured Application: This proof-of-concept work might find applications as programmable zoom systems for imaging systems or holographic projection systems, when applied with newer high resolution SLMs.

Abstract: In this work we present an experimental proof of concept of a programmable optical zoom lens system with no moving parts that can form images with both positive and negative magnifications. Our system uses two programmable liquid crystal spatial light modulators to form the lenses composing the zoom system. The results included show that images can be formed with both positive and negative magnifications. Experimental results match the theory. We discuss the size limitations of this system caused by the limited spatial resolution and discuss how newer devices would shrink the size of the system.

Keywords: zoom lens; spatial light modulators; imaging systems

---

1. Introduction

Optical zoom systems are basic components of all optical imaging systems. These zoom systems change the magnification of the image without altering the position of the object and image planes. They are usually composed of lens systems, in which a mechanical displacement of the lenses varies the effective magnification of the system, while keeping fixed object and image planes [1].

The design of zoom lenses without mechanical movements has been proposed since the early production of tunable lenses [2]. Programmable active lenses are currently available with different technologies, including liquid lenses [3], liquid-crystal lenses [4,5], or active polymer lenses [6]. Pancharatnam–Berry (PB) lenses are a new technology useful to overcome the slow response time of other technologies [7]. Therefore, zoom lenses with tunable adjustment and no moving elements have been reported [8–11].

Phase-only spatial light modulators (SLMs) are also very useful devices to display a lens function. They can display lenses where the focal length can be controlled directly by the geometry of the pattern addressed from a computer. They are diffractive lenses, and therefore they are affected by chromatic dispersion. They also require the use of polarized light. In addition, the pixelated structure of the SLMs impose limits on the shortest focal length that can be implemented. Nevertheless, SLMs allow a great flexibility to combine multiple lens functions [12], to multiplex the lens function with other holographic functions [13], or to correct aberrations [14]. These functions cannot be implemented with other kinds of tunable lenses.

Zoom lenses composed by SLMs have been reported previously [15]. In ref. [16], an anamorphic zoom lens system was demonstrated that used two SLMs. In Refs. [17,18], a holographic zoom lens projector using SLMs was also produced. However, SLMs present certain limitations and artifacts that make lenses displayed on them differ from the perfect idealized paraxial tunable lens. Nonlinear phase modulation, quantization of the phase levels, amplitude and/or polarization modulation coupled to phase modulation, fringing effects, phase fluctuation (flicker), or non-flatness panels that introduce aberrations, are some of the typical phenomena that one can find in these devices [13]. All these effects deteriorate the optical performance of lenses displayed onto SLM and must be considered and reduced, when possible.

In this work, we discuss the limits imposed by the rasterized structure of the SLM and give details for its experimental implementation in a simple zoom system composed of two lenses. We present configurations, which allow for different magnifications, and also to change the sign of the magnification to obtain both erect and inverted images. Because we employ SLMs with relatively large pixel sizes, we are forced to build a very large optical system, impractical for realistic applications. However, the results here presented constitute an experimental proof-of-concept that can be made useful and compact by using the newer SLM technology.

2. Geometry of the System

Figure 1 shows the configuration of our optical system. A programmable SLM is placed a distance $p = 100$ cm from an input object. The object (a slide transparency with the word "PROFESSOR") is illuminated by a collimated beam from a helium-neon laser, with the $\lambda = 633$ nm wavelength. A second SLM is placed a distance $d = 140$ cm from the first. Finally, the camera is located a distance also approximately $q = 100$ cm from the second SLM. The output patterns were recorded with a CCD camera, model XC-37 (Sony, USA), having 491×384 pixels on a 8.86×6.60 mm^2 sensor.

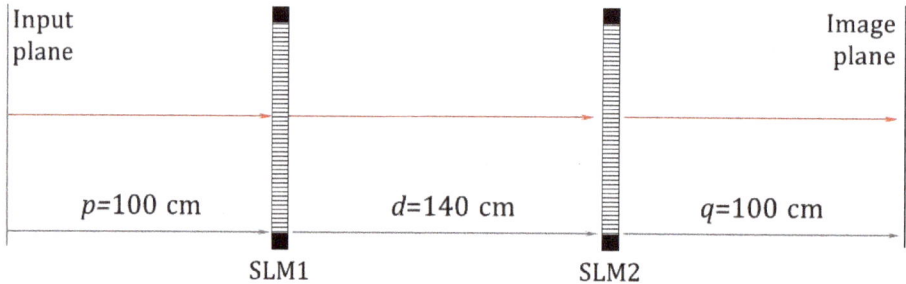

Figure 1. Schematic of the zoom lens optical system.

Note that zoom lenses used in photography are designed to image distant objects, so p is a very large distance. Tunable zoom lenses have also been demonstrated for holographic projection systems [17,18]. There, the p is a short distance, while q should be larger. Here we selected an intermediate approach where the object and image are located at finite close distances.

Using a standard ray matrix approach [19], this system can be represented by the product of three translation matrices and the two lens matrices as in Equation (1). The product of these matrices is given by the final ABCD matrix:

$$\begin{vmatrix} r_2 \\ r_2' \end{vmatrix} = \begin{vmatrix} 1 & q \\ 0 & 1 \end{vmatrix} \times \begin{vmatrix} 1 & 0 \\ -1/f_2 & 1 \end{vmatrix} \times \begin{vmatrix} 1 & d \\ 0 & 1 \end{vmatrix} \times \begin{vmatrix} 1 & 0 \\ -1/f_1 & 1 \end{vmatrix} \times \begin{vmatrix} 1 & p \\ 0 & 1 \end{vmatrix} \times \begin{vmatrix} r_1 \\ r_1' \end{vmatrix} = \begin{vmatrix} A & B \\ C & D \end{vmatrix} \times \begin{vmatrix} r_1 \\ r_1' \end{vmatrix} \quad (1)$$

r_i and r_i' ($i = 1, 2$) denote the ray height and angle coordinates.

The calculation of the ABCD matrix in Equation (1) leads to the following result:

$$
\begin{vmatrix} r_2 \\ r_2' \end{vmatrix} = \begin{vmatrix} 1 - \frac{d}{f_1} - \frac{q}{F} & p\left(1 - \frac{d}{f_1} - \frac{q}{F}\right) + d + q\left(1 - \frac{d}{f_2}\right) \\ -\frac{1}{F} & -\frac{p}{F} + 1 - \frac{d}{f_2} \end{vmatrix} \times \begin{vmatrix} r_1 \\ r_1' \end{vmatrix}
\tag{2}
$$

where F denotes the focal length of the composed system, given by:

$$
\frac{1}{F} = \frac{1}{f_1} + \frac{1}{f_2} - \frac{d}{f_1 f_2}
\tag{3}
$$

The imaging requirement is met by forcing the matrix element $B = 0$. This condition leads to the following relation

$$
f_2 = \frac{f_1 q(d+p) - pdq}{f_1(p+d+q) - p(d+q)} = \frac{f_1 q(D-q) - pdq}{f_1 D - p(D-p)}
\tag{4}
$$

where $D = p + d + q$ is the total distance of the system, which we want to keep constant, so the system has no moving elements.

The magnification is given by the matrix element A in Equation (2), i.e.,

$$
m = 1 - \frac{d+q}{f_1} - \frac{q}{f_2} - \frac{qd}{f_1 f_2}
\tag{5}
$$

The system is limited by the geometry of the SLMs. In this work, the first lens is encoded onto a CRL model XGA-3 transmissive twisted nematic SLM operating in a phase-only mode by means of the rotated eigen-polarization state technique [20]. This device has 1024 × 768 pixels with a pixel spacing of 18 microns. The second lens is formed on a Hamamatsu model X10468-01 reflective parallel-aligned SLM, also operating in a phase-only mode. This device has 800 × 600 pixels with a pixel spacing of 20 microns. Because this second SLM is a reflective liquid-crystal on silicon (LCoS) display, we include a beam splitter in the system to allow a reflective geometry in the second part of the system in Figure 1. Both displays use eight bits for codification of phase levels, i.e., they have $N = 256$ quantization levels. The first order diffraction efficiency caused by the quantization of phase levels is given by $\eta_Q = [\sin(\pi/N)/(\pi/N)]^2$ which for $N = 256$ levels results in a value of exceeding 99% [21]. Thus, quantization is not an issue for this system. Both SLMs have been tested and produce a phase-only response versus addressed gray level, with no cross-modulation effects. In addition, the phase modulation versus addressed gray level is linearized.

On the contrary, the limited spatial resolution imposes drastic losses and limits. Although both devices have similar pixel pitch, the LCoS technology presents a much higher fill factor (FF). The fill factor is defined as FF = $(w/\Delta)2$, where w stands for the pixel width and Δ stands for the pixel spacing [22]. The CRL display has a fill factor of approximately only $FF_{CRL} \approx 35\%$. In contrast, the Hamamatsu LCoS display presents a much higher value of approximately $FF_{HAM} \approx 98\%$. This difference makes a major impact between the two devices. First, the fraction of the incident intensity that is transmitted is given directly by the FF. Secondly, the SLM rasterized structure created a 2D grid of diffraction orders. The central principal zero order has the highest intensity, which is roughly proportional to FF^2 [22]. Since we are using the zero diffracted order in the zoom lens system, the total intensity transmission is reduced by this effect to a factor of $\eta_{FF} = (FF_{CRL} \times FF_{HAM})^2 \approx 12\%$.

But, even more relevant for the application to the zoom lens system is the limitation imposed by the limited spatial resolution on the minimum focal length that the SLMs can encode. We have previously discussed [13] the limits imposed by the spatial resolution of the SLM when we implement a lens function having a quadratic phase dependence as:

$$
Z^*(r, f) = \exp\left(-i\frac{\pi r^2}{\lambda f}\right)
\tag{6}
$$

where f denotes the focal length of the lens, λ is the wavelength and r denotes the radial coordinate at the SLM plane. Next, we summarize the main effect, for its implication in the zoom lens system.

We assume that the SLM has an array of $N \times N$ pixels where each pixel has a finite size given by Δ. As the focal length f of the encoded lens decreases, the spacing between the phase at the edges of the device can become smaller than the Nyquist limit of 2 pixels causing aliasing effects. These considerations lead to the Nyquist focal length (f_N) for a lens encoded onto the SLM as given by [12]:

$$f_N = \frac{N\Delta^2}{\lambda} \tag{7}$$

Lens functions as in Equation (6) with focal lengths shorter (in absolute value) than the Nyquist focal length $|f| \leq f_N$ are affected by aliasing and will not be well reproduced.

Figure 2 shows three diffractive lenses where the focal lengths are greater and smaller than the Nyquist limit. If the focal length is clearly greater than f_N, the lens function is encoded perfectly. However, for shorter focal lengths the effective central area of the lens decreases and the lens begins to act as a low pass spatial filter. In addition, the strengths of the replica focal lengths caused by the aliasing increase.

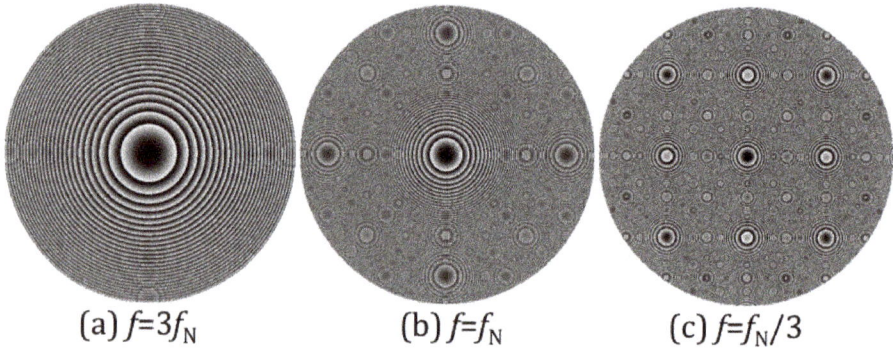

Figure 2. Phase patterns for diffractive lenses having focal lengths of (**a**) $3f_N$, (**b**) f_N and (**c**) $f_N/3$.

For our SLMs, the effective Nyquist focal lengths are about 38 cm, considering the smaller dimensions. These considerations have important consequences for our optical system because the SLMs have difficulty encoding focal lengths smaller than this value. This limit requires the large values for the distances in our experimental system in Figure 1. These large dimensions of the zoom lens system make it impractical for real applications. However, we show results as discussed next that demonstrate the proof-of-concept and illustrate the potential usefulness of newer high resolution SLMs to build programmable zoom lens systems

Although the proper characterization of a zoom lens system requires the evaluation of parameters like resolution, field of view, modulation transfer function or contrast, we concentrate in these proof-of-concept results only in the different magnifications that the system can produce including a change of sign.

3. Experimental Imaging Results without Moving Elements

As stated earlier, the object is an opaque slide with the transparent word "PROFESSOR" having a size of roughly 9×1 mm. Figure 3 shows the image when the camera is placed directly against the slide, so the reader can compare the object size with the images formed for different configurations of the zoom system. Note that the image is not perfectly in focus due to the small propagation distance between the slide and the camera detector.

Figure 3. Input object captured by camera. Here the slide with the word "PROFESSOR" is placed directly to the detector. The size of the object slightly exceeds the size of the detector.

3.1. Negative Magnifications

We first examine cases where we obtain inverted images. Figure 4 presents the system configuration and the output image when the first SLM is encoded with a focal length of $f_1 = 70.2$ cm and the second SLM is turned off. Therefore, we have a single lens imaging system, where the object distance is p, and the image distance is $d + q$. The output is inverted with a magnification of about $m = -2.3$ in excellent agreement with the theoretical magnification obtained using Equation (5). The size of the image is larger than the detector. As we moved the position of the input slide, different areas of the letter were seen. We note that, in doing these experiments, we simply varied the focal length encoded onto the device until the image was focused. This procedure accounts for small deviations of the distances p, d and q.

Figure 5 shows the zoom system configuration and the output image when the first SLM is turned off and the second SLM is encoded with a focal length of $f_2 = 69.6$ cm. Again, we have a single lens imaging system, where now the image distance q is smaller than the object distance $p + d$. The output is inverted with a magnification of about $m = -0.43$ in excellent agreement with the theoretical magnification obtained by using Equation (5). The size of the image is smaller than the detector.

Finally, Figure 6 shows the output image when both devices are encoded with focal lengths of $f_1 = f_2 = 100$ cm. Now the focal length of the composite lens system is approximately $F = 167$ cm, and the object is inverted with a magnification of $m = -1$ in excellent agreement with theory. Note that this configuration places the input object plane in the front focal plane of the first lens and the final output image plane in the back focal plane of the second lens.

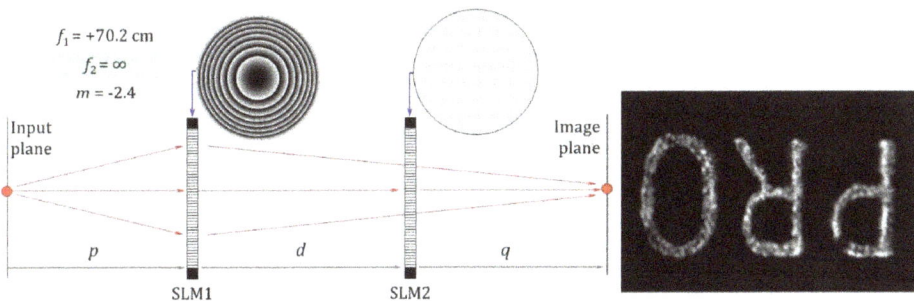

Figure 4. Zoom system configuration with $f_1 = 70.2$ cm and $f_2 = \infty$ and experimental image with magnification of $m = -2.4$.

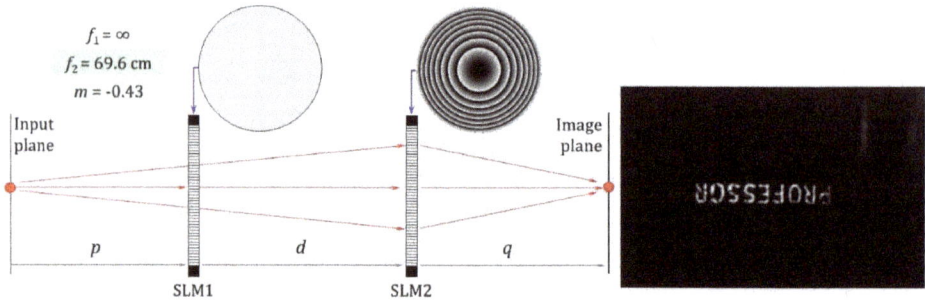

Figure 5. Zoom system configuration with $f_1 = \infty$ and $f_2 = +69.6$ cm and experimental image with magnification of $m = -0.4$.

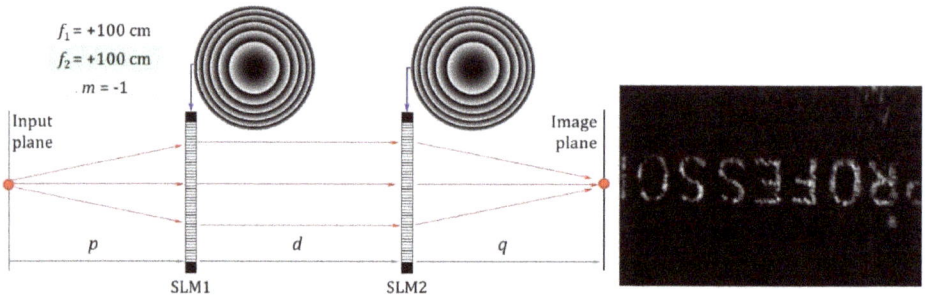

Figure 6. Zoom system configuration with $f_1 = +100$ cm and $f_2 = +100$ cm and experimental image with magnification of $m = -1$.

Note that in these three cases shown here, the encoded lenses have focal lengths larger than the Nyquist limit of 38 cm. Therefore, the displayed lenses are well reproduced and the final image is well focused in all cases. There is, however, some low pass filtering caused by the limited size of the SLMs compared to the long distances required in the system.

3.2. Positive Magnifications

In this section, we examine positive magnifications where the image has the same orientation as the object. These results have not been previously presented, to our knowledge. In each case, the first lens forms a real image between the two lenses that is inverted and serves as a real object for the second lens. This second lens then forms the erect image.

Figure 7 shows the case where the first SLM is encoded with a focal length of $f_1 = 50$ cm while the second SLM is encoded with a focal length of $f_2 = 31.8$ cm. Note that this second focal length is below the Nyquist limit. The composed system has a negative focal length of about $F = -27$ cm. We now obtain an erect image with a magnification of $m = +2$ in excellent agreement with the theoretical value. The size of the image is larger than the detector.

Figure 8 shows the case where the first SLM is encoded with a focal length of $f_1 = 31.8$ cm while the second SLM is encoded with a focal length of $f_2 = 50$ cm. Now the first lens has a focal length shorter than the Nyquist limit. The composed system again has a negative focal length of about $F = -27$ cm. Again we obtain an erect image, but with a smaller magnification of $m = +0.5$ in excellent agreement with theory. We cannot capture the entire image, unlike the case in Figure 5. In this case because of the short focal length of the first lens, the rays forming the external points of the intermediate image do not enter the second lens. This effect is similar to that shown in the holographic zoom system in ref. [17]. As we moved the object, other parts of the image would come into view. This effect is also present in Figure 7, but is not as visible because of the large magnification such that the image is larger than the area of the camera.

Finally, Figure 9 shows the case where the first SLM is encoded with a focal length of $f_1 = 40$ cm while the second SLM is encoded with a focal length of $f_2 = 44$ cm. The composed system again has a negative focal length of about $F = -31$ cm. Again we obtain an erect image, but with a magnification of $m = +1.0$ in excellent agreement with theory. Again, we cannot capture the entire image because of the short focal lengths required.

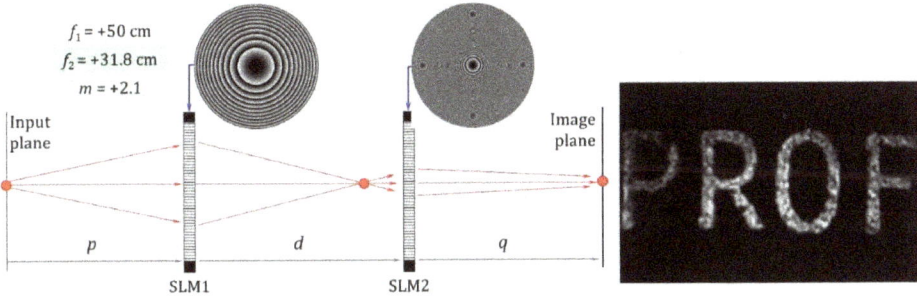

Figure 7. Zoom system configuration with $f_1 = +50$ cm and $f_2 = +31.8$ cm and experimental image with magnification of $m = +2$.

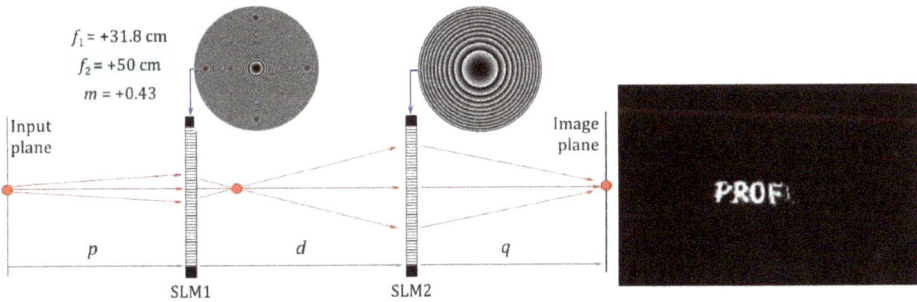

Figure 8. Zoom system configuration with $f_1 = +31.8$ cm and $f_2 = +50$ cm and experimental image with magnification of $m = +0.5$.

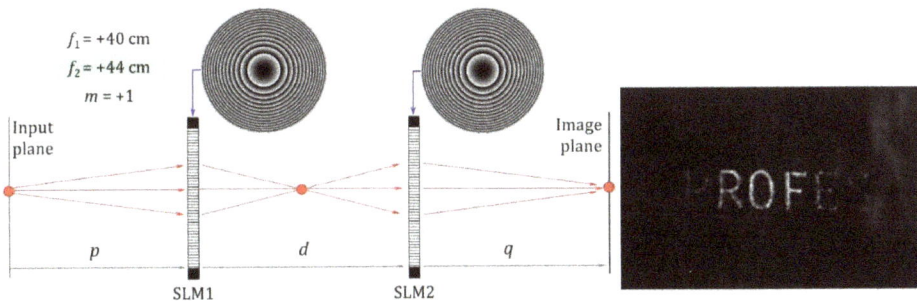

Figure 9. Zoom system configuration with $f_1 = +40$ cm and $f_2 = +44$ cm and experimental image with magnification of $m = +1$.

4. Conclusions

In conclusion, we demonstrate a programmable zoom lens system where the magnification and sense of the image can be controlled without moving any parts. We use two programmable SLMs onto which we can encode the required focal length lenses to achieve these results. We note that the physical size of this system is extremely large primarily because of the small sizes of the lenses (diameters of 14 mm and 12 mm respectively) and the Nyquist limits on the focal lengths that we can encode onto our SLMs.

However, there are new high resolution 4KUHD(Ultra-High Definition) SLM models having a pixel size as small as 3.74 microns in an array of 3840×2160 pixels (data from the Exulus-4K1 model [23] offered by Thorlabs), or 4160×2464 (data from the the GAEA-2 model [24], offered by Holoeye). The pixel spacing of these SLMs lead to a Nyquist focal length limit of about 50 mm. These devices would allow a much more compact zoom lens system.

Another strategy for reducing the size of the system is to combine the lenses displayed on the SLMs with other glass lenses that add a constant optical power. However, for simplicity, here we analyze the system where all the optical power is encoded only on the SLMs.

Nevertheless, we obtain excellent results and show a proof-of-concept of a zoom system without moving elements based on SLMs, and showing negative magnifications ranging from -2 to $-1/2$ and negative magnifications from $+2$ to $+1/2$. The use of SLMs to produce the zoom system can be very promising since other functionalities of SLMs like wavefront aberration correction, optical processing functions, or polarization control can be incorporated into the system.

Author Contributions: Conceptualization, J.A.D. and I.M.; Methodology, J.A.D., I.M. and D.M.C.; Software, D.M.C.; Validation, J.A.D., T.I.H., and J.P.S.; All authors; Writing-Review & Editing, J.A.D.; Supervision, J.A.D.

Funding: The participation of I.M. in this investigation was funded by Ministerio de Economía y Competitividad from Spain and FEDER Funds (ref. FIS2015-66328-C3-3-R) and from Conselleria d'Educació, Investigació, Cultura i Esport, Generalitat Valenciana (ref. PROMETEO-2017-154).

Conflicts of Interest: The authors declare no conflict of interest.

References

1. Youngworth, R.N.; Betensky, E.I. Fundamental considerations for zoom lens design. *Proc. SPIE* **2012**, *8488*, 848806.
2. Tam, E.C. Smart electro-optical zoom lens. *Opt. Lett.* **1992**, *17*, 369–371. [CrossRef] [PubMed]
3. Kuiper, S.; Hendriks, B.H.W. Variable-focus liquid lens for miniature cameras. *Appl. Phys. Lett.* **2004**, *85*, 1128–1130. [CrossRef]
4. Ren, H.; Fox, D.W.; Wu, B.; Wu, S.T. Liquid crystal lens with large focal length tunability and low operating voltage. *Opt. Express* **2007**, *15*, 11328–11335. [CrossRef] [PubMed]
5. Jamali, A.; Bryant, D.; Zhang, Y.; Grunnet-Jepsen, A.; Bhowmik, A.; Bos, P. Design of a large aperture tunable refractive Fresnel liquid crystal lens. *Appl. Opt.* **2018**, *57*, B10–B19. [CrossRef] [PubMed]
6. Yun, S.; Park, S.; Nam, S.; Park, B.; Park, S.K.; Mun, S.; Lim, J.M.; Kyung, K.-U. An electro-active polymer based lens module for dynamically varying focal system. *Appl. Phys. Lett.* **2016**, *109*, 141908. [CrossRef]
7. Lee, Y.-H.; Tan, G.; Zhan, T.; Weng, Y.; Liu, G.; Gou, F.; Peng, F.; Tabiryan, N.V.; Gauza, S.; Wu, S.-T. Recent progress in Pancharatnam–Berry phase optical elements and the applications for virtual/augmented realities. *Opt. Data Process. Storage* **2017**, *3*, 79–88. [CrossRef]
8. Santiago, F.; Bagwell, B.E.; Pinon, V.; Krishna, S. Adaptive polymer lens for rapid zoom shortwave infrared imaging applications. *Opt. Eng.* **2014**, *53*, 125101. [CrossRef]
9. Li, H.; Cheng, X.; Hao, Q. An electrically tunable zoom system using liquid lenses. *Sensors* **2016**, *16*, 45. [CrossRef] [PubMed]
10. Li, L.; Yuan, R.-Y.; Wang, J.-H.; Wang, Q.-H. Electrically optofluidic zoom system with a large zoom range and high-resolution image. *Opt. Express* **2017**, *25*, 22280–22297. [CrossRef] [PubMed]
11. Kopp, D.; Brender, T.; Zappe, H. All-liquid dual-lens optofluidic zoom system. *Appl. Opt.* **2017**, *56*, 3758–3763. [CrossRef] [PubMed]

12. Cottrell, D.M.; Davis, J.A.; Hedman, T.R.; Lilly, R.A. Multiple imaging phase-encoded optical elements written as programmable spatial light modulators. *Appl. Opt.* **1990**, *29*, 2505–2509. [CrossRef] [PubMed]

13. Haist, T.; Osten, W. Holography using pixelated spatial light modulators—Part 1: Theory and basic considerations. *J. Micro/Nanolithogr. MEMS MOEMS* **2015**, *14*, 041310. [CrossRef]

14. Neil, M.A.A.; Booth, M.J.; Wilson, T. Closed-loop aberration correction by use of a modal Zernike wave-front sensor. *Opt. Lett.* **2000**, *25*, 1083–1085. [CrossRef] [PubMed]

15. Wick, D.V.; Martinez, T. Adaptive optical zoom. *Opt. Eng.* **2004**, *43*, 8–9.

16. Iemmi, C.; Campos, J. Anamorphic zoom system based on liquid crystal displays. *J. Eur. Opt. Soc. Rapid Pub.* **2009**, *4*, 09029. [CrossRef]

17. Lin, H.-C.; Collings, N.; Chen, M.-S.; Lin, Y.-H. A holographic projection system with an electrically tuning and continuously adjustable optical zoom. *Opt. Express* **2012**, *20*, 27222–27229. [CrossRef] [PubMed]

18. Chen, M.-S.; Collings, N.; Lin, H.-C.; Lin, Y.-H. A holographic projection system with an electrically adjustable optical zoom and a fixed location of zeroth-order diffraction. *J. Disp. Technol.* **2014**, *10*, 450–455. [CrossRef]

19. Davis, J.A.; Lilly, R.A. Ray-matrix approach for diffractive optics. *Appl. Opt.* **1993**, *32*, 155–158. [CrossRef] [PubMed]

20. Davis, J.A.; Moreno, I.; Tsai, P. Polarization eigenstates for twisted-nematic liquid crystal displays. *Appl. Opt.* **1998**, *37*, 937–945. [CrossRef] [PubMed]

21. Moreno, I.; Iemmi, C.; Márquez, A.; Campos, J.; Yzuel, M.J. Modulation light efficiency of diffractive lenses displayed onto a restricted phase-mostly modulation display. *Appl. Opt.* **2004**, *43*, 6278–6284. [CrossRef] [PubMed]

22. Davis, J.A.; Chambers, J.B.; Slovick, B.A.; Moreno, I. Wavelength dependent diffraction patterns from a liquid crystal display. *Appl. Opt.* **2008**, *47*, 4375–4380. [CrossRef] [PubMed]

23. Exulus Spatial Light Modulator. Available online: https://www.thorlabs.com/newgrouppage9.cfm?objectgroup_id=10378 (accessed on 19 June 2018).

24. GAEA-2 Megapixel Phase Only Spatial Light Modulator (Refective). Available online: https://holoeye.com/spatial-light-modulators/gaea-4k-phase-only-spatial-light-modulator/ (accessed on 19 June 2018).

applied
sciences

MDPI

Article

Anamorphic and Local Characterization of a Holographic Data Storage System with a Liquid-Crystal on Silicon Microdisplay as Data Pager

Fco. Javier Martínez-Guardiola [1,2,*], Andrés Márquez [1,2], Eva M. Calzado [1,2], Sergio Bleda [1,2], Sergi Gallego [1,2], Inmaculada Pascual [2,3] and Augusto Beléndez [1,2]

[1] Department of Physics, Systems Engineering and Signal Theory, Universidad de Alicante, 03690 Alicante, Spain; andres.marquez@ua.es (A.M.); evace@ua.es (E.M.C.); sergio.bleda@ua.es (S.B.); sergi.gallego@ua.es (S.G.); a.belendez@ua.es (A.B.)
[2] Instituto Universitario de Física Aplicada a las Ciencias y las Tecnologías, Universidad de Alicante, 03690 Alicante, Spain; pascual@ua.es
[3] Department of Optics, Pharmacology and Anatomy, Universidad de Alicante, 03690 Alicante, Spain
* Correspondence: fj.martinez@ua.es; Tel.: +34-965-903-692

Received: 8 May 2018; Accepted: 11 June 2018; Published: 15 June 2018

Abstract: In this paper, we present a method to characterize a complete optical Holographic Data Storage System (HDSS), where we identify the elements that limit the capacity to register and restore the information introduced by means of a Liquid Cristal on Silicon (LCoS) microdisplay as the data pager. In the literature, it has been shown that LCoS exhibits an anamorphic and frequency dependent effect when periodic optical elements are addressed to LCoS microdisplays in diffractive optics applications. We tested whether this effect is still relevant in the application to HDSS, where non-periodic binary elements are applied, as it is the case in binary data pages codified by Binary Intensity Modulation (BIM). To test the limits in storage data density and in spatial bandwidth of the HDSS, we used anamorphic patterns with different resolutions. We analyzed the performance of the microdisplay in situ using figures of merit adapted to HDSS. A local characterization across the aperture of the system was also demonstrated with our proposed methodology, which results in an estimation of the illumination uniformity and the contrast generated by the LCoS. We show the extent of the increase in the Bit Error Rate (BER) when introducing a photopolymer as the recording material, thus all the important elements in a HDSS are considered in the characterization methodology demonstrated in this paper.

Keywords: holographic data storage; holographic and volume memories; parallel-aligned; liquid-crystal on silicon; liquid crystals; spatial light modulator; photopolymer

1. Introduction

Since the first laser developments, Holographic Data Storage Systems (HDSS) have been a promising and appealing technology for true 3D storage of information and associative memory retrieval [1,2]. There are some scientific and technological challenges to fulfill when a HDSS is developed [3,4].

Our group evaluated the introduction of some novelties in the HDSS. One of these is the introduction of a parallel-aligned liquid crystal on silicon (PA-LCoS) microdisplay [5]. The PA-LCoS acts as data pager, and we designed some modulation schemes for it [6]. In this paper, we present a characterization method that allows us to evaluate the limitations introduced by different elements in our optical system.

In previous works, we developed a characterization method for PA-LCoS, enabling to characterize the flicker effects produced because of the digital addressing technology, and also to characterize the retardance introduced [7]. In the literature, other degradation phenomena have also been detected and analyzed related not only with modern PA-LCoS microdisplays but also with previous liquid crystal displays (LCD). An important one is the anamorphic and frequency dependent effect [8–11] that causes a reduction in the performance depending on the spatial frequency and on the orientation of the image displayed. In this sense, the anamorphic frequency dependent effect is exhibited by liquid crystal displays (LCDs) [8,9,11] and by modern LCoS devices, both digital [10] and analogically [11] driven. In LCDs, it was demonstrated [9] that the anamorphic phenomenon was directly related with the electronics driving the device. In general, the electrical signal corresponding to an image addressed to the display is produced by multiplexing the rows composing the image, where each row contains the voltage value to be applied to each of the pixels in the row. As a result, the horizontal (or pixel) frequency in the electrical signal is much larger than the vertical (or row) frequency. The limited bandwidth of the electronics driver will then produce a low-pass filtering for the horizontal frequency components in the image, especially if the image contains fast variations along the horizontal. This low-pass filtering of the electrical signal is then reflected on the reduction of the phase modulation range available when displaying phase-only diffractive optical elements [8]. The anamorphic and frequency dependent phenomenon has also been reported when using the LCD in the amplitude-only regime [3,12–15]. The authors were interested on the characterization of the modulation transfer function (MTF) of the LCD for its application in optical processing [12–14]. In [3] (p. 247) and [15], the LCD is applied in holographic data storage, which is also the focus of the present paper.

Besides the effects produced by the limited electronics bandwidth already described, interpixel effects in the LC layer may also cause a reduction in the spatial frequency bandwidth. These interpixel effects are the fringing-field (i.e., appearance of tangential components in the applied electric field) and vicinity LC adherence effects that appear as a result of the competition between the tangential components of the applied field and the intermolecular forces within the liquid crystals (i.e., avoiding abrupt changes in the orientation) [16–22]. These interpixel effects are becoming more and more important in modern LCoS devices as the pixel size is becoming increasingly small, even smaller than 4 μm [22], thus decreasing the ratio between the pixel size and the LC layer thickness (i.e., the cell gap). We also note that vicinity LC adherence effects exhibit an anamorphic behavior, as described in numerical simulations by Wang et al. [18], depending on whether the applied electric voltage gradient is along the LC director at the alignment layers or perpendicular.

Our first goal in this paper is to identify if the anamorphic and frequency dependent phenomenon affects our particular application, which in the present case is the application of PA-LCoS microdisplays to holographic data storage. We also need to test other limitations such as the maximum resolution of the optical system, aberrations, etc. We want to know the origin of these limitations to change the setup to improve the performance and to know our practical data storage capacity.

In this paper, we focus in the complete optical system. Thus, we evaluate the limitations due to the lenses, the ones added by the PA-LCoS microdisplay, and the illumination. All this work is necessary prior to the introduction of the holographic material, which adds additional degradation effects. The methodology that we propose is not only able to provide a global characterization of the HDSS, but also to perform a local and quantitative characterization across the aperture of the system. This local characterization is very useful to evaluate nonuniformity of the illumination, spatial variant response of the LCoS device, and misalignments of the elements in the system.

As we are managing a binary data storage system, to study the influence of the different elements, we measured the Bit Error Rate (BER) or the Quality factor (Q-factor), figures of merit defined for digital data transmission systems. We analyzed the errors retrieved in the reconstructed image by means of a CCD camera.

We show how the introduction of the material affects the performance. We used Polyvinyl Alcohol Acrylamide (PVA/AA) that was produced, characterized and modeled by our group [23]. These preliminary results show us how to improve the process of producing and depositing PVA/AA for holographic data storage applications.

2. Experimental Setup

In our HDSS, we used a convergent processor architecture for the object beam. This setup is also known as VanderLugt correlator [24]. This alternative provided us some flexibility. It allowed a wider freedom when selecting lenses to be used, because it is not necessary that the focal lengths match, as is the case in the classical 4-f system. Furthermore, the areal density on the recording material can be varied without the need to change the lenses, as explained in [25].

In Figure 1, we show the complete experimental setup. The object beam follows the path formed by lenses L2 to L5. Along the same optical axis, we have lens L8 to form the image of the data page onto the CCD camera plane. Path formed by lenses L6 and L7 is the reference beam. In our discussion about limitations, we only refer to the object beam combined with lens L8.

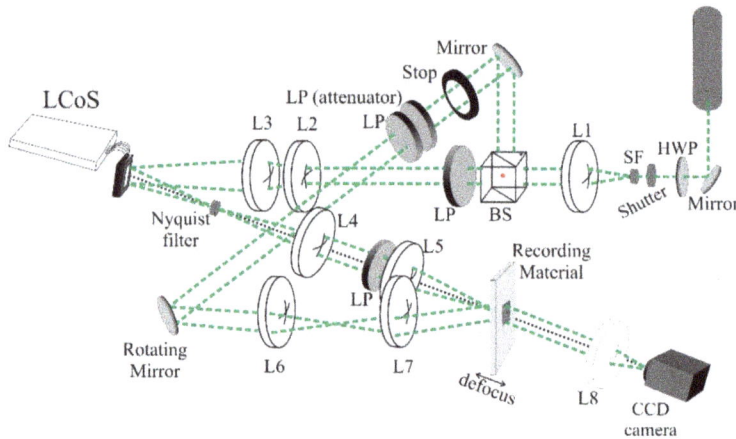

Figure 1. Diagram of the experimental setup using convergent processor.

In the object beam, we introduced the data page by means of the PA-LCoS. The display was placed after the L3 lens. We used a divergent lens L2, so that the combination of L2 and L3 enables to modify the convergence angle of the beam onto the LCoS. The beam impinges onto the PA-LCoS while it is converging. The Fourier transform forms at the convergence plane, where we placed a Nyquist filter to control the Fourier Transform orders that we store. Lenses L4 and L5 form a relay system to image the Nyquist filter plane, which contains our information, onto the recording plane. In the present work, we did not use the filtering option by the Nyquist filter, since our focus was the study of the anamorphic and optical limitations of our system.

The PA-LCoS used in this work is a commercially available PA-LCoS microdisplay, model PLUTO distributed by the company HOLOEYE. It is filled with a nematic liquid crystal, with 1920 × 1080 pixels and 0.7" diagonal. The fill factor is 87% and its pixel pitch is 8.0 μm. It is important to remark the pixel pitch, because it defines the information bit size for different resolutions. The incident angle to separate impinging and reflecting object beam is 11.5°, and it is fixed during all the experiment.

We used different numbers of pixels in the PA-LCoS to form a bit of information. These bit sizes were: 8 × 8, 4 × 4, 3 × 3, and 2 × 2 pixels. This means that, for every bit of information, we used, respectively, 64, 16, 9 and 4 pixels in the PLUTO device. Thus, they have, respectively, a 64 μm, 32 μm, 24 μm, and 16 μm length in each side for the square that forms the information bit. In this work, we aggregated the bits in the shape of linear stripes, arranged in a vertical or horizontal orientation, because we wanted to test if the reduction in the performance of the HDSS system is related to an anamorphic effect in the PA-LCoS. In this context, the bit size defines the minimum stripe width.

In the data analysis, for statistical reasons, we considered individual bits of information, instead of individual stripes, to calculate the Bit Error Rate: in this way, we ensured that we have enough information to statistically analyze the image. We addressed LCoS data pages with 512×512 pixels. If we would consider stripes as a block of information, we would only have 64 information blocks for 8×8 pix. size, 128 for 4×4, etc. Considering individual bits, we have 4096 bits of information for 8×8 pix. size, 16,384 for 4×4, etc.

To retrieve information, we used a CCD camera. We used the PCO.1600 model from PCO.imaging company. This is a high dynamic 14-bit cooled CCD camera with a resolution of 1600×1200 pixels, and a pixel size of 7.4×7.4 μm^2. The magnification of the PA-LCoS plane onto the camera plane is about a factor of two. Therefore, there is no limitation from the CCD camera since the image built on the CCD plane is oversampled.

3. Characterization Method

When analyzing the data storage system, we find various sources of limitations. The initial one is produced by the aperture of the lenses. For that reason, we have to test the resolution capability of our setup by using a standard 1951 USAF test target. So, the first step for characterizing the optical system is to obtain the resolving power of the complete optical system composed of lenses L4, L5 and L8. To do that, we relay the image from the PA-LCoS plane onto the CCD camera but we place a negative 1951 USAF resolution test chart instead of our PA-LCoS device, and we capture the reflected image with the help of the CCD camera. Reference beam is blocked and no recording material is introduced.

In Figure 2, we show the retrieved image of the 1951 USAF resolution test target. From this image, we conclude that our limit is Group 4 element 6 for both horizontal and vertical orientation, which corresponds to a resolution of 28.50 lp/mm or equivalently 35.09 μm per line pair. This a line width of 17.54 μm, which, translated into pixels in our PA-LCoS, means that a stripe width of two pixels (8 + 8 μm) cannot be resolved.

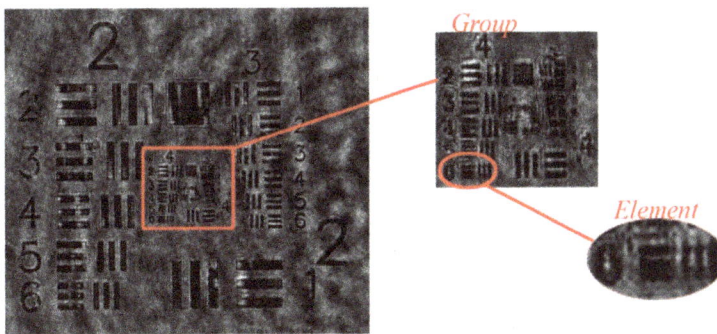

Figure 2. Relayed image when the negative USAF pattern is used instead of PA-LCoS.

Once we calculated our optical system resolution limit, we introduced the PA-LCoS instead of the USAF resolution test target. We evaluated whether there were additional limitations related with the well-known anamorphic effects presented by this kind of microdisplays [10,11,22]. To do that, in our optical system, we prepared some test patterns with different resolutions and with two orientations, vertical and horizontal.

The images have the aspect presented in Figure 3, where we show the pattern that corresponds with a bit size of 8×8 pixels in the PA-LCoS. We alternated between black and white stripes of a selected width. The position and width of the stripeswere generated in a random way to avoid periodic diffractive effects. We added fiducial marks for centering and rotation, and to detect the position of the information bits. These marks also enabled evaluating the quality of the image or distortion during the capture. The width of the lines composing the fiducial marks was always 8 pixels, which was our reference to evaluate the resolution used in the data page in the case we do not know it.

We configured our setup for binary intensity modulation (BIM) [6]. To do that, we uploaded to the PA-LCoS an electrical configuration that offers a 180° retardance range, with a good linearity, and permitted obtaining the maximum contrast and less flicker, as we analyzed in [26]. The maximum contrast would be produced when the polarizers in the object beam are crossed with respect to each other and they form a 45° angle with respect to the alignment direction of the LC director in the PA-LCoS display.

(a)	(b)

Figure 3. Vertical and Horizontal patterns used; (**a**): Vertical stripes pattern; (**b**): Horizontal stripes pattern.

In Figure 4, we show the calibration measurements done for the PA-LCoS. These calibration measurements were obtained with the configuration described above, and the 11.5° incident angle commented in the setup description. As we used BIM, we only used two values of gray level. These values have to offer the maximum contrast. In Figure 4, we selected gray level 14 as black level (lowest intensity) and gray level 248 as white level (highest intensity).

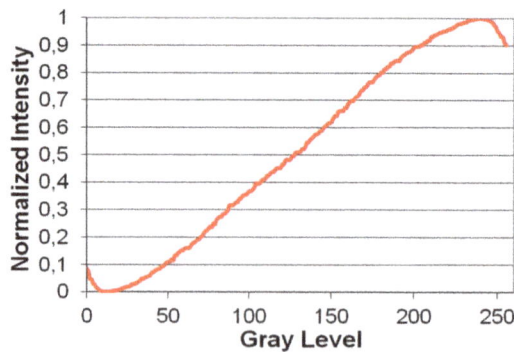

Figure 4. Normalized intensity introduced by the PA-LCoS.

To analyze the errors introduced by the optical system, we illuminated the PA-LCoS (see Figure 1) and, through lenses L4, L5 and L8, we produced the image of the data page onto the CCD camera [25]. The image data obtained was then digitally analyzed [6]. We note that positions for lenses L4, L5 and L8 are the same as in the previous characterization step with the USAF resolution test target and there is no recording material in the system.

After capturing the image with the CCD camera and before processing the image to convert it into a binary image, we extracted the region of the image that contains the data, i.e., the stripes region. To detect this region, we used fiducial marks. These marks allowed calculating the scale factor introduced, the rotation, and the possible distortion effects produced by the optical setup. The rotation can be corrected by software to prepare the image to extract the data.

We applied some processing operations in the captured image to retrieve the information stored. These processing operations were also applied when we proceeded, afterwards, with the characterization of the complete system in conjunction with recording material. Since we used a binary image, we selected the best threshold value to distinguish between the 0 level and the 1 level. For that reason, we thresholded the image by selecting the appropriate value of the gray level to separate the 0s from the 1s. We analyzed various values to select the level that minimizes the detected errors, and for every gray level we calculate the BER. The CCD has a bit depth of 14 bits, which means that we have values from 0 to 16384.

One possibility is to calculate the BER by counting directly the falsely detected bits in the image. To this goal, we first applied a thresholding. In this way, we obtained a binary image. Then, we compared the bits in the binary image obtained with the bits in the original data set displayed in the PA-LCoS, which must to be done for every threshold gray level considered. In the thresholded image, the bit is composed of a block of CCD pixels. We considered the value of the central CCD pixel as the value for the bit. By counting the number of errors, we could calculate the BER. This BER was calculated by dividing the number of errors by the total number of bits. The total number of bits depends on the resolution selected, which means that for a 8 × 8 bit size we have 64 × 64 (4096) bits of information, 128 × 128 (16,384) bits for 4 × 4 bit size, 170 × 170 (28,900) bits for 3 × 3 bit size and 256 × 256 (65,536) bits for a 2 × 2 bit size.

This form of calculating the BER can be useful as a first approximation but it depends too much on the specific image captured. Therefore, it is better to calculate the Q-factor, which offers an estimation measure for the quality of the signal-to-noise ratio. This Q-factor is calculated from the probability distribution of 1s and 0s from the unprocessed images. As long as we can infer the positions of the 1s and 0s, we can obtain a histogram of 1s and 0s and calculate this Q-factor that it is typically used in digital systems in fiber optics communications [27–29]. Q-factor is given by,

$$Q = \frac{|\mu_1 - \mu_0|}{\sigma_1 + \sigma_0} \tag{1}$$

where μ_1 and μ_0 are the mean value in the histograms produced for the gray level distribution of ON and OFF bits, respectively, and σ_1 and σ_0 are the corresponding standard deviations. If the histograms are approximated as Gaussian distributions, the relation between BER and Q-factor is given by,

$$BER = \frac{1}{2}\left[1 - erf\left(\frac{Q}{\sqrt{2}}\right)\right] \tag{2}$$

where *erf* is the error function [27–29], which is tabulated in various mathematical handbooks [30,31]. These values show the capacity of the optical system to recover the digital information introduced. We needed a raw BER in the range of 10^{-3} to assure a complete information recovery after introducing error correction codes.

In Figure 5, we show a thresholded image captured for a data page with a pixel size of 8 × 8 (8 × 8 pixels in the PA-LCoS for each bit of information). Figure 5 also shows the division of the image in bits, given by the overlain grid pattern. We present this image to illustrate that, even though the images proposed in the study only have information in a specific direction (horizontal or vertical), to allow a statistical analysis, we analyzed the image at a bit level.

Figure 5 represents the best thresholded image detected for an 8 × 8 data page; to make the bits more visible, we have used the biggest bit size. These 8 × 8 pixels per bit are referred to the image displayed in the PA-LCoS. The image shown in Figure 5 is the image captured by the CCD: in this case, every bit of information is formed by 16 × 16 CCD pixels, because of the lens image magnification (about ×2) and considering the different pixel size in the PA-LCoS and in the CCD. In the image, we easily identify the wrong CCD pixels within the error bits (encircled in Figure 5).

To calculate the Q-factor in Equation (1), which permits us a more reliable BER calculation, we needed more statistical information. To use this figure of merit, we divided into bits the original image, before applying any threshold on it, and we associated the gray level with the position of an ON or an OFF state (1 or 0). Then, we took the value measured by the CCD and countrf the number of bits at that level. Figure 6 shows the distribution of the different gray levels where we have identified the ON and OFF states. We see that there is a zone where the histograms of levels ON and OFF overlap. This zone contains the best level to threshold the image to minimize the number or errors. The extension of the zone, where the states overlap, also defines the number of errors that we encounter in the analysis.

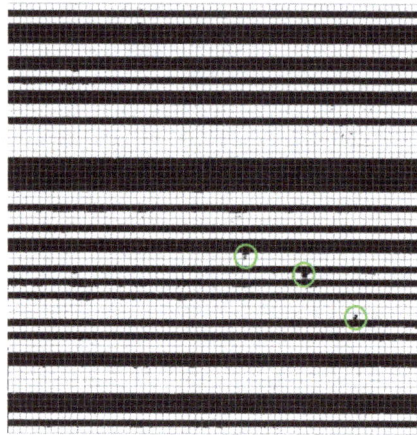

Figure 5. Thresholded image with a grid that indicates the position of the bits and possible CCD pixel errors marked.

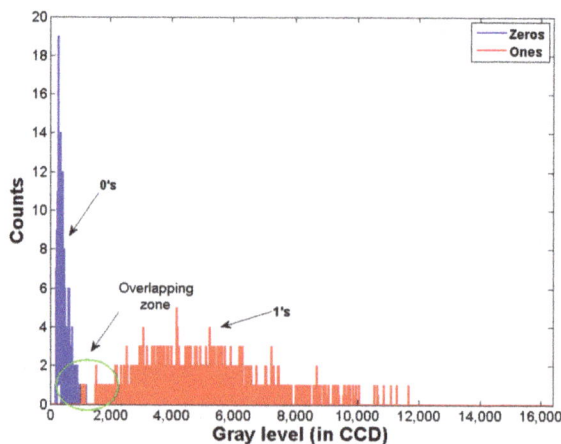

Figure 6. Histogram of ON and OFF states.

4. Results

In this section, we present the results for the different images obtained. As we have mentioned, we used bit aggregates in the shape of linear stripes oriented vertically and horizontally to test the anamorphic effects. In addition, we changed the resolution in the image displayed: we used as information unit a group of pixels in the PA-LCoS. In Figure 7, we plot the images captured by the CCD camera when addressing the stripe data shape onto the PA-LCoS, respectively, for horizontal and vertical orientation for a bit size of 4 × 4 pixels. These are the original images before the thresholding operation.

The first thing to notice is the irregular illumination observed. This produces and image where a unique threshold level for the entire image does not exist. As we can vary the threshold level to make an image binarization, we can obtain some figures of merit to analyze the irregular illumination observed, and we can calculate the number of errors for all the threshold levels considered.

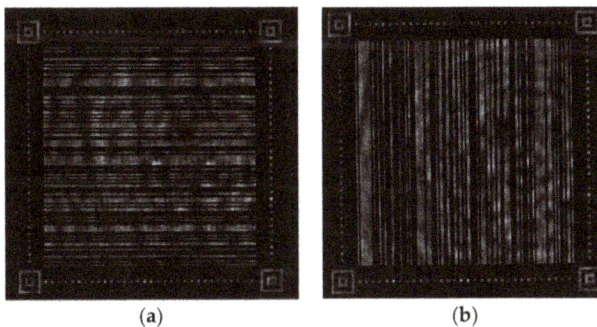

(a) (b)

Figure 7. Original images captured for a 4 × 4 pixel resolution horizontal and vertical orientations, previous to thresholding; (**a**): Captured image for horizontal stripes pattern (**b**): Captured image for vertical stripes pattern.

The images depicted in Figure 7 are the ones captured by our CCD camera with a bit depth of 14 bits, so we have 16,384 possible gray levels. As long as we avoid saturating the CCD, we do not need to consider a full variation of all levels when we try to seek for the best threshold level. It is usually enough to consider a variation of 6000 levels, which is more than the typical extent of the overlapping zone (see Figure 6).

Figure 8 presents the evolution in the number of errors counted as a function of the threshold level. The graph width gives us an idea about how sensitive is the information reconstruction to the threshold level. In fact, we observe how the number of errors decreases to a minimum and then rises again. The minimum level of light is not zero, since there always exists a thermal noise or a residual light intensity, thus it will be enough to start analyzing from a threshold level greater than 100, in our case we consider 200. The best threshold level would be the one that minimizes the number of errors. To be sure that we have selected the best threshold, we have to obtain a curve with an evolution, as the one reflected in either Figure 8a,b, which means that the number of error decreases until a minimum point and then the number of error rises again. That shape means that we have selected a range large enough to contain the threshold level that minimizes the number or errors.

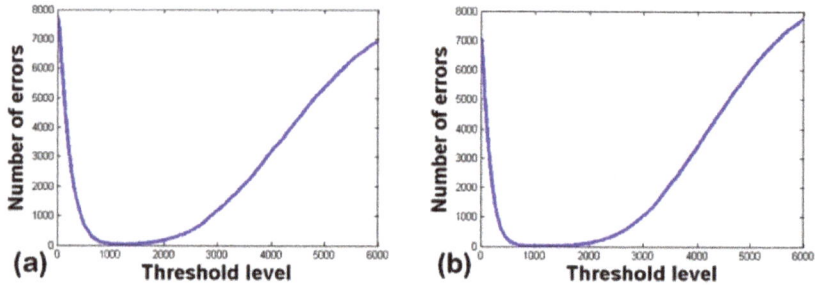

Figure 8. Number of errors counted as a function of threshold level for a complete image: (**a**) horizontal orientation; and (**b**) vertical orientation.

To analyze the illumination influence, we now divide the image into four zones. Each of them can be analyzed in the same way described for the full image. In Figure 9, we show how we have defined the zones for the different orientations. Each zone is a different set of data to extract statistical results, as presented in the previous section.

Figure 10 shows the different graphs for the zones defined in Figure 9 for vertical stripes. These zones have the same shape, but we can observe that they have slightly different best threshold levels for every zone. As long as the graph shape is preserved, the differences do not produce a significant increase in the number of errors.

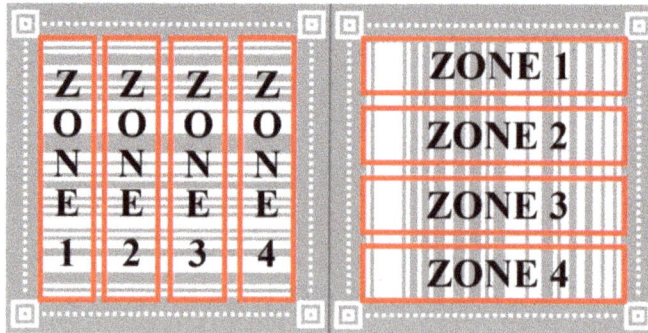

Figure 9. Zones defined for uniformity analysis.

To maintain the possibility to compute statistical results, we do not reduce the size of the zones to look for illumination inhomogeneities. Reducing the size of the zones decreases the number of information bits available to calculate histograms. Nevertheless, as a qualitative evaluation tool, we have calculated the number of errors for every column (horizontal stripes) or row (vertical stripes). In this way, we see that there are columns (or rows) in which we identify all 0s and 1s correctly. This kind of graphs allows us to study and evaluate the illumination, but it is not statistically significant to calculate Q-factor or BER per row (or column), because we only have 128 bits of data for each column or row in an image that use 4×4 pixels per bit. In the best case, we would have 512 bits per column when using 1×1 pixel per bit.

In Figure 11, the number of errors is pseudocolored: blue represents few errors than red. We show the number of errors detected in a specific column in Figure 11a (row in Figure 11b). The column number is represented in the X-axis (when we use vertical stripes, this number refers to a specific row), and the evolution in the number of errors as a function of the threshold level is represented in the Y-axis. This representation gives us some clues about the illumination and how sensitive is the threshold level selection. A wider blue zone, which is correlated with the large valley in Figure 8, implies that the selection of the threshold level is less critical. The ripples shown in Figure 11 come from the illumination inhomogeneities: we can see that the horizontal orientation (Figure 11a) is more sensitive to these inhomogeneities.

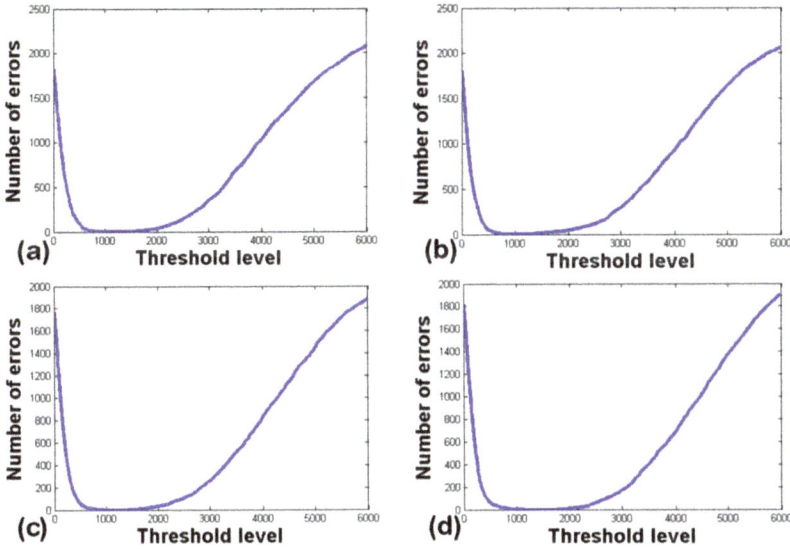

Figure 10. Number of errors counted as a function of threshold level for vertical stripes orientation by zone: (**a**) Zone 1; (**b**) Zone 2; (**c**) Zone 3; and (**d**) Zone 4.

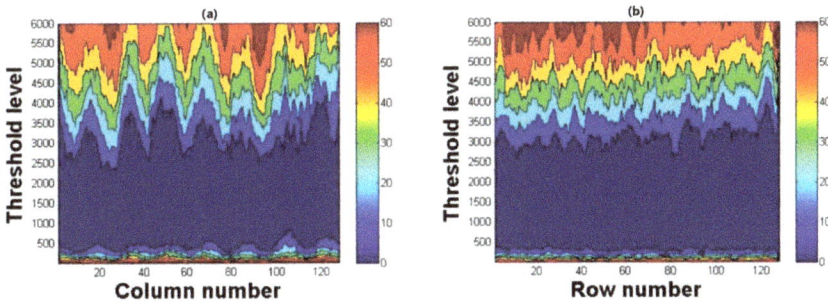

Figure 11. Number of errors for each column (horizontal stripes) and row (vertical stripes) as a function of threshold level. Data page with 4 × 4 pixels per bit. (**a**) number of errors detected in a specific column; (**b**) number of errors detected in a specific row.

For the different bit sizes and linear stripes orientations, we provide specific quantitative results in tabular form. Tables 1 and 2 shows the data page as a whole, and Tables 3 and 4 shows them in terms of the local characterization in four regions across the aperture of the data page.

Table 1. Bit error rate and Q-factor for Horizontal stripes.

Pixel Size	Q-Factor	Errors	BER
8 × 8	3.082	1	1.0×10^{-3}
4 × 4	2.473	25	6.8×10^{-3}
3 × 3	2.243	117	12.5×10^{-3}
2 × 2	0.574	18,097	2.8×10^{-1}

Table 2. Bit error rate and Q-factor for Vertical stripes.

Pixel Size	Q-Factor	Errors	BER
8 × 8	3.124	3	9.0×10^{-4}
4 × 4	2.884	8	2.0×10^{-3}
3 × 3	2.650	26	4.0×10^{-3}
2 × 2	1.699	2371	4.5×10^{-2}

Table 3. Measured Q-factor by zone and for Horizontal stripes.

Full Image	Zone 1	Zone 2	Zone 3	Zone 4
		8 × 8 pixels bit size		
3.082	3.266	3.185	2.977	3.110
		4 × 4 pixels bit size		
2.473	2.516	2.648	2.411	2.462
		3 × 3 pixels bit size		
2.243	2.226	2.354	2.202	2.297

Table 4. Measured Q-factor by zone and for Vertical stripes.

Full Image	Zone 1	Zone 2	Zone 3	Zone 4
		8 × 8 pixels bit size		
3.124	3.558	3.439	2.955	2.909
		4 × 4 pixels bit size		
2.884	3.195	3.012	2.790	2.685
		3 × 3 pixels bit size		
2.650	2.668	2.586	2.747	2.671

First, in Tables 1 and 2, we show the measured Q-factor and BER for the different bit sizes and orientations considered. The Q-factor is calculated using Equation (1), and the data were obtained from the histograms (Figures 8 and 10). The BER was obtained applying Equation (2). We have also included the number of errors, counting directly, associated with the optimum threshold level. The first thing that we note is that the bit size of 2 × 2 pixels presents a very high BER and a low Q-factor. This is because this bit size is in the resolution limit of our system, as we calculated from the 1951 USAF resolution test target shown in Figure 2. Therefore, this resolution is out of our scope due to the limitations imposed by the optical system, not to the LCoS display. Table 1 also reflect that the BER is in the order of 10^{-3} or below before introducing error correction codes. This fact indicates that the reconstruction without errors is possible [28,29].

From the results in Tables 1 and 2, we also see that the Q-factor and BER values are better, i.e., less image degradation, for the vertical stripes images. This contradicts the results in diffractive optical elements displayed on LCoS devices [8,10,11] where diffraction efficiency becomes smaller for vertically oriented elements, i.e., periodicity along the columns of the LCoS. One possible explanation for this difference is that in the HDSS application in the paper, where BIM data pages are introduced, there is wider tolerance to the anamorphic effects introduced by LCoS devices. Now, we do not display periodic diffractive elements and the figure of merit is not diffraction efficiency. In our present case, the binary intensity contrast between the ON and OFF levels in the LCoS is the important magnitude and this has a limited dependency with the orientation of the non-periodic stripes displayed.

We divided the data image into various zones, as presented in Figure 9, to analyze the Q factor and the BER by zones. This kind of analysis helps us to detect problems in the uniformity of the illumination, or errors due to deformations in the image.

In Tables 3 and 4, we show the Q-factor obtained in the different zones. In horizontal stripes (Table 3), Zone 1 corresponds to the left part on the image. We observe that, for every bit size, the worst Q-factor is obtained in Zone 3. Nevertheless, for vertical stripes (Table 4), we see that the Q-factor decreases with the number of the zone, that means from top to the bottom (Figure 9, left). This kind of analysis allows us to measure the illumination inhomogeneity: if the difference is too high, we would have to improve the experimental setup to avoid this source of error.

We present results for the complete reconstruction process. When we introduce the photopolymer, new distortion elements appear. The manufacturing process can introduce ripples and deformations due to continuous exchange of water molecules between material and environment. In addition, the natural crystallization may affect the recording process [32].

In Figure 12, we show the image captured by the CCD during the reconstruction process. We show the reconstructed images for the 8 × 8 and 4 × 4 bit sizes to compare with the results obtained without material. The degradation effects increase the BER, and lower the Q-factor accordingly, as can be observed in Tables 5 and 6, for the analysis of the reconstructed data pages with the horizontal (Table 5) and vertical patterns (Table 6) with the recording material.

Table 5. BER and Q-factor for Horizontal stripes in the reconstructed images.

Pixel Size	Q-Factor	Errors	BER
8 × 8	1.995	46	23×10^{-3}
4 × 4	1.514	850	65×10^{-3}

Table 6. BER and Q-factor for Vertical stripes in the reconstructed images.

Pixel Size	Q-Factor	Errors	BER
8 × 8	1.687	109	46×10^{-3}
4 × 4	1.663	486	48×10^{-3}

Comparing the results in Tables 5 and 6, when the material is introduced, with the corresponding results in Tables 1 and 2, we see that the number of errors increases by one order of magnitude, which makes the resolutions under a pixel size of 4 × 4 out of the values that can be recovered with an error correction code. Perhaps, the non-uniformity illumination is influencing in the registration process, but it is even more likely that slight non-uniformity in the deposition of the material has a relevant role in the degradation of the BER.

We expect that the BER obtained for a single hologram will not be affected by multiplexing as long as the Bragg selectivity (in the case of angular multiplexing) is maintained, as long as we do not saturate the dynamic range of the material and maintain an appropriate relation between diffraction efficiency and the number of holograms stored [33].

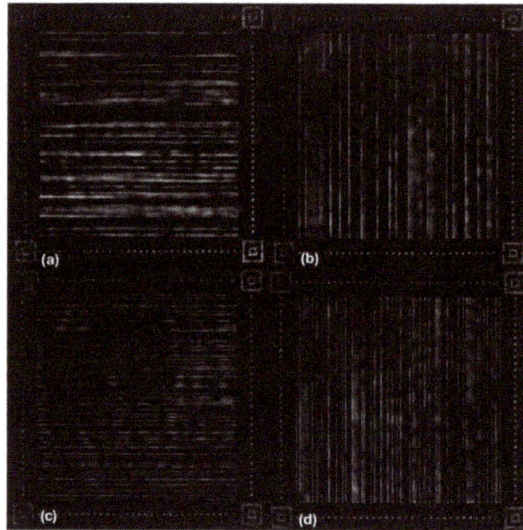

Figure 12. Reconstructed images for 8 × 8 (upper row) and 4 × 4 (lower row) pixels stripes. Horizontal (left column) and Vertical (right column). (**a**) 8 × 8, horizontal; (**b**) 8 × 8, vertical; (**c**) 4 × 4, horizontal; (**d**) 4 × 4, vertical.

5. Conclusions

In this paper, we present a method to discover the limits of an Holographic Data Storage System, with an emphasis on the existence of anamorphic and frequency dependent effects due the PA-LCoS microdisplay. We defined two figures of merit to evaluate the possibilities of doing an exact reconstruction: Q-factor and BER.

We discovered that the anamorphic effect does not influence the results when a non-periodic pattern is used. This effect seems to have an influence when we used periodic diffractive optical elements [34]. We evaluated two figures of merit, BER and Q-factor, exposed in Equations (1) and (2), which serve to predict that, without using recording material, we can reconstruct data pages with a resolution of 3 × 3 pixels per bit in a reliable way. The fact that the BER is in the order of 10^{-3} or below indicates that our HDSS can be used to store and retrieve data with a high data density.

The different graphs that we obtain with this characterization method also allow us to see illumination problems that have to be solved to improve the performance. To do that, we divided the image into four zones. This local evaluation can be done without adding elements to the experimental setup. We can also extract qualitative information about illumination by analyzing the image column by column (or row by row). We consider that the method is easy to implement and useful to detect limitations and problems in Holographic Data Storage Systems in a systematic way.

We analyzed the BER when the holographic material is introduced and, despite the increased number of errors, the BER is good enough to consider that the data can be recovered. However, we need to improve the homogeneity in the material deposition, or even try to use a commercial material such as Bayfol®, which is made industrially and is protected with a film that prevents material degradation and guaranties uniformity [35].

Author Contributions: F.J.M.-G. and A.M. conceived and designed the experiments, adjusted the experimental setup and wrote the article. F.J.M.-G. and E.M.C. performed the experiments and collected the images. F.J.M.-G. and S.B. analyzed the data and performed statistical analysis. I.P. and A.B. assisted in the experimental setup, part of the data analysis, and revising the manuscript. E.M.C. and S.G. manufactured of the photopolymer used in the experiments.

Acknowledgments: This was supported by Ministerio de Economía, Industria y Competitividad (Spain) under projects FIS2017-82919-R (MINECO/AEI/FEDER, UE) and FIS2015-66570-P (MINECO/FEDER), Generalitat Valenciana (Spain) under project PROMETEO II/2015/015 and Universidad de Alicante (Spain) under project GRE17-06.

Conflicts of Interest: The authors declare no conflict of interest.

References

1. Van Heerden, P.J. Theory of optical information storage in solids. *Appl. Opt.* **1963**, *2*, 393–400. [CrossRef]
2. Sarid, D.; Schechtman, B.H. A roadmap for data storage applications. *Opt. Photonics News* **2007**, *18*, 32–37. [CrossRef]
3. Coufal, H.J.; Psaltis, D.; Sincerbox, B.T. (Eds.) *Holographic Data Storage*; Springer: New York, NY, USA, 2000. [CrossRef]
4. Curtis, K.; Dhar, L.; Hill, A.; Wilson, W.; Ayres, M. (Eds.) *Holographic Data Storage: From Theory to Practical Systems*; John Wiley & Sons, Ltd.: Chichester, UK, 2010.
5. Lazarev, G.; Hermerschmidt, A.; Kruger, S.; Osten, S. *LCoS Spatial Light Modulators: Trends and Applications, in Optical Imaging and Metrology: Advanced Technologies*; Osten, W., Reingand, N., Eds.; Wiley-VCH Verlag & Co.: Weinheim, Germany, 2012.
6. Martínez, F.J.; Fernández, R.; Márquez, A.; Gallego, S.; Álvarez, M.L.; Pascual, I.; Beléndez, A. Exploring binary and ternary modulations on a PA-LCoS device for holographic data storage in a PVA/AA photopolymer. *Opt. Express* **2015**, *23*, 20460–20479. [CrossRef] [PubMed]
7. Martínez, F.J.; Márquez, A.; Gallego, S.; Francés, J.; Beléndez, A.; Pascual, I. Retardance and flicker modeling and characterization of electro-optic linear retarders by averaged Stokes polarimetry. *Opt. Lett.* **2014**, *39*, 1011–1014. [CrossRef] [PubMed]
8. Márquez, A.; Iemmi, C.; Moreno, I.; Campos, J.; Yzuel, M.J. Anamorphic and spatial frequency dependent phase modulation on liquid crystal displays. Optimization of the modulation diffraction efficiency. *Opt. Express* **2005**, *13*, 2111–2120. [CrossRef] [PubMed]
9. Márquez, A.; Moreno, I.; Iemmi, C.; Campos, J.; Yzuel, M.J. Electrical origin and compensation for two sources of degradation of the spatial frequency response exhibited by liquid crystal displays. *Opt. Eng.* **2007**, *46*, 114001. [CrossRef]
10. Lobato, L.; Lizana, A.; Márquez, A.; Moreno, I.; Iemmi, C.; Campos, J.; Yzuel, M.J. Characterization of the anamorphic and spatial frequency dependent phenomenon in liquid crystal on silicon displays. *J. Eur. Opt. Soc. Rapid Publ.* **2011**, *6*, 11012S. [CrossRef]
11. Albero, J.; García-Martínez, P.; Martínez, J.L.; Moreno, I. Second order diffractive optical elements in a spatial light modulator with large phase dynamic range. *Opt. Lasers Eng.* **2013**, *51*, 111–115. [CrossRef]
12. Hsieh, M.L.; Hsu, K.Y.; Paek, E.G.; Wilson, C.L. Modulation transfer function of a liquid crystal spatial light modulator. *Opt. Commun.* **1999**, *170*, 221–227. [CrossRef]
13. Hsieh, M.L.; Paek, E.G.; Wilson, C.L.; Hsu, K.Y. Performance enhancement of a joint transform correlator using the directionality of a spatial light modulator. *Opt. Eng.* **1999**, *38*, 2118–2121. [CrossRef]
14. Grother, P.; Casasent, D. Modulation transfer function measurement method for electrically addressed spatial light modulators. *Appl. Opt.* **2001**, *40*, 5253–5259. [CrossRef] [PubMed]
15. Márquez, A.; Gallego, S.; Méndez, D.; Álvarez, M.L.; Fernández, E.; Ortuño, M.; Neipp, C.; Beléndez, A.; Pascual, I. Accurate control of a liquid-crystal display to produce a homogenized Fourier transform for holographic memories. *Opt. Lett.* **2007**, *32*, 2511–2513. [CrossRef] [PubMed]
16. Apter, B.; Efron, U.; Bahat-Treidel, E. On the fringing-field effect in liquid-crystal beam-steering devices. *Appl. Opt.* **2004**, *43*, 11–19. [CrossRef] [PubMed]
17. Efron, U.; Apter, B.; Bahat-Treidel, E. Fringing-field effect in liquid-crystal beam-steering devices: An approximate analytical model. *J. Opt. Soc. Am.* **2004**, *21*, 1996–2008. [CrossRef]
18. Wang, X.; Wang, B.; Bos, P.J.; McManamon, P.F.; Pouch, J.J.; Miranda, F.A.; Anderson, J.E. Modeling and design of an optimized liquid-crystal optical phased array. *J. Appl. Phys.* **2005**, *98*, 073101. [CrossRef]
19. Persson, M.; Engström, D.; Goksör, M. Reducing the effect of pixel crosstalk in phase only spatial light modulators. *Opt. Express* **2012**, *20*, 22334–22343. [CrossRef] [PubMed]

20. Lingel, C.; Haist, T.; Osten, W. Optimizing the diffraction efficiency of SLM-based holography with respect to the fringing field effect. *Appl. Opt.* **2013**, *52*, 6877–6883. [CrossRef] [PubMed]

21. Lu, T.; Pivnenko, M.; Robertson, B.; Chu, D. Pixel-level fringing-effect model to describe the phase profile and diffraction efficiency of a liquid crystal on silicon device. *Appl. Opt.* **2015**, *54*, 5903–5910. [CrossRef] [PubMed]

22. Wang, M.; Martínez, F.J.; Márquez, A.; Ye, Y.; Zong, L.; Pascual, I.; Beléndez, A. Polarimetric and diffractive evaluation of 3.74 micron pixel-size LCoS in the telecommunications C-band. In Proceedings of the SPIE 10395, Optics and Photonics for Information Processing XI, 103951J, San Diego, CA, USA, 24 August 2017. [CrossRef]

23. Gallego, S.; Ortuño, M.; Neipp, C.; Fernández, E.; Beléndez, A.; Pascual, I. Improved maximum uniformity and capacity of multiple holograms recorded in absorbent photopolymers. *Opt. Express* **2007**, *15*, 9308–9319. [CrossRef] [PubMed]

24. VanderLugt, A. *Optical Signal Processing*; John Wiley & Sons: Chichester, UK, 1992.

25. Márquez, A.; Fernández, E.; Martínez, F.J.; Gallego, S.; Ortuño, M.; Beléndez, A.; Pascual, I. Analysis of the geometry of a holographic memory setup. In Proceedings of the SPIE 8429, Optical Modelling and Design II, 84291Y, Brussels, Belgium, 4 May 2012.

26. Martínez, F.J.; Márquez, A.; Gallego, S.; Ortuño, M.; Francés, J.; Beléndez, A.; Pascual, I. Electrical dependencies of optical modulation capabilities in digitally addressed parallel aligned liquid crystal on silicon devices. *Opt. Eng.* **2014**, *53*, 067104. [CrossRef]

27. Ramamoorthy, L.; Kumar, V.K.; Hoskins, A.; Curtis, K. *Data Channel Modeling, Chapter 10. Holographic Data Storage: From Theory to Practical Systems*; John Wiley & Sons, Ltd.: Chichester, UK, 2010; pp. 221–245.

28. Agrawal, G.P. *Fiber-Optic Communication Systems*; John Wiley & Sons, Inc.: Chichester, UK, 2010. [CrossRef]

29. Keiser, G. *Optical Fiber Communications*, 4th ed.; McGraw-Hill: New York, NY, USA, 2011.

30. Navidi, W. *Principles of Statistics for Engineers and Scientists*; McGraw-Hill: New York, NY, USA, 2010.

31. Zwillinger, D. (Ed.) *Standard Mathematical Tables and Formulae*, 31st ed.; CRC Press: Boca Raton, FL, USA, 2003.

32. Ortuño, M.; Gallego, S.; García, C.; Neipp, C.; Beléndez, A.; Pascual, I. Optimization of a 1mm thick PVA/acrylamide recording material to obtain holographic memories: Method of preparation and holographic properties. *Appl. Phys. B* **2003**, *76*, 851–857. [CrossRef]

33. Mok, F.H.; Burr, G.W.; Psaltis, D. System metric for holographic memory systems. *Opt. Lett.* **1996**, *21*, 896–898. [CrossRef] [PubMed]

34. Martínez, F.J.; Márquez, A.; Gallego, S.; Ortuño, M.; Francés, J.; Pascual, I.; Beléndez, A. Predictive capability of average Stokes polarimetry for simulation of phase multilevel elements onto LCoS devices. *Appl. Opt.* **2015**, *54*, 1379–1386. [CrossRef] [PubMed]

35. Bruder, F.-K.; Fäcke, T.; Rölle, T. The Chemistry and Physics of Bayfol® HX Film Holographic Photopolymer. *Polymers* **2017**, *9*, 472. [CrossRef]

applied sciences

MDPI

Article

Transmission Matrix Measurement of Multimode Optical Fibers by Mode-Selective Excitation Using One Spatial Light Modulator

Stefan Rothe *, Hannes Radner, Nektarios Koukourakis and Jürgen W. Czarske

Technische Universität Dresden, Laboratory of Measurement and Sensor System Technique, Helmholtzstraße 18, 01069 Dresden, Germany; hannes.radner@tu-dresden.de (H.R.); nektarios.koukourakis@tu-dresden.de (N.K.); juergen.czarske@tu-dresden.de (J.W.C.)
* Correspondence: stefan.rothe@tu-dresden.de

Received: 14 December 2018; Accepted: 4 January 2019; Published: 8 January 2019

Abstract: Multimode fibers (MMF) are promising candidates to increase the data rate while reducing the space required for optical fiber networks. However, their use is hampered by mode mixing and other effects, leading to speckled output patterns. This can be overcome by measuring the transmission matrix (TM) of a multimode fiber. In this contribution, a mode-selective excitation of complex amplitudes is performed with only one phase-only spatial light modulator. The light field propagating through the fiber is measured holographically and is analyzed by a rapid decomposition method. This technique requires a small amount of measurements N, which corresponds to the degree of freedom of the fiber. The TM determines the amplitude and phase relationships of the modes, which allows us to understand the mode scrambling processes in the MMF and can be used for mode division multiplexing.

Keywords: multimode fiber; digital holography; mode division multiplexing; transmission matrix

1. Introduction

Fiber optical networks are the internet infrastructure's backbone. In recent decades, the exponentially increasing data rate has been managed by exploiting multiplexing techniques on singlemode fibers (SMF). Following this trend, the rising amount of sensitive data is commonly protected by cryptographic algorithms, that have a common flaw—they can be decoded with infinite computation power. Hence, alternative techniques have to be investigated.

One possible approach to enhance the information security in fiber optical networks is to use MMFs where, by reason of production tolerances in fiber manufacturing or deformation, excited modes in the wave guide become significantly scrambled during propagation. Occurring scattering processes in MMFs have already been compensated [1–3], but are now investigated as an advantage considering information security.

One possibility to increase potential transmission rates is to use multiple single- or few-mode fiber cores taken together in one fiber bundle [4]. However, by usage of one optical multi mode transmission system, Mode Division Multiplexing (MDM) is applicable, which is predicted as the most effective technique to replace multiple communication channels at once, outperforming the transmission rates of SMF-based networks [5–7]. MDM was developed to address space as the fourth dimension following time, polarization and wavelength to send information [8].

Knowledge of the TM of the MMF is required for both the information security and MDM in order to recover the channels by optical means. On one hand, the TM provides essential information on the level of secrecy and on the other hand on performance characterization concerning the MMF as a multi-mode device. The level of secrecy is measured by the uncertainty of an eavesdropper about a sent message [9]. For these contexts, the TM describes the amplitude and phase relations between the spatial and polarization modes of the complete fiber's mode domain.

Several groups have already presented approaches to obtain access to the TM of disordered media like MMFs. In [10], the first measurement of a TM has been published, where the light propagation through a random scattering sample was investigated. By using a phase-only spatial light modulator (SLM), Hadamard patterns, whose elements are either +1 or −1 in amplitude, were chosen as the input basis for the illumination of the sample. The light that propagated through the sample was interferometrically measured using phase-shifting holography (PSH), which produces a measurement effort of at least $4N$ (N = number of controlled segments). The presented work paved the way for various studies based on a TM observation. However, the explained procedure was developed for a sample having generalised scattering properties, but in terms of MMF applications, there are more practical ways to acquire the TM. Aiming to use the MMF as a flexible image transmission system, for example, as an ultra-thin endoscope, the TM is used to reconstruct a desired image at one end of the fiber [11,12]. For this reason, focal spots scanning the input fiber facet of the MMF are generated with an SLM. The output signal is again measured using PSH. The mode-based TM is then generated by determining a conversion matrix that transfers the focal spot domain to the fiber's mode domain. Concerning information-technical approaches in this case, it is more accurate to relinquish the approximation step of determining a conversion matrix and to directly excite the modes in the MMF to observe the coupling process. In [7,13], the supported MMF modes are sequentially excited by using an SLM. Additionally, another SLM is applied at the fiber output in order to decompose the MMF output signal into the mode domain. A correlation filter method is used, where the efficiency of diffracted light off the SLM surface into a power meter is proportional to the correlation of the fiber output with a desired mode. Consequently, the TM entries are obtained individually and the measurement effort is N^2. Likewise, the presented technique in [14] to obtain the TM distinguishes a holographic setup at the fiber output. They measure the pure intensity distributions at the MMF output and use a computational algorithm to determine the individual TM entries. However, optimization approaches like the introduced *complex semi-definite programming* have a major flaw concerning the computational effort in the case of measuring TMs with a large number of modes.

In this paper, an optical setup is developed to acquire the fiber's TM by using an SLM at the fiber input to sequentially excite every mode that the waveguide supports. The light's phase and amplitude is directly shaped using proper combinations of neighbored SLM pixels forming a superpixel. Since the selective mode excitation is challenging to align, a thorough instruction of the presented setup is given. The light field that propagated through the MMF is analyzed using a novel approach. The MMF output is recorded with digital holography (DH) and the amplitude and phase of the captured signal are reconstructed numerically with the *angular spectrum* method obtaining the complex light field distribution. The measured field distribution is used to determine the individual TM entries by decomposing it into the fiber's mode domain. The presented procedure reduces the measurement effort to N, which means that one row of a TM is captured single-shot.

2. Procedure of Operation

TM measurements are commonly performed using an SLM that displays different masks to imprint a desired field distribution on an incident Gaussian beam [10–14]. For a few-mode fiber, inaccuracies in mode generation are still manageable, since deviations are not likely to overlap with any other mode, because there are only a few other modes in the fiber. For this reason, the power that does not overlap with the desired mode simply leads to a loss rather than an unwanted mode excitation [7]. MMFs are significantly more sensitive to undesired mode excitation. As more modes are in the MMF's mode domain, it is more probable that power that does not overlap with the desired mode is overlapping with another. For this reason, it is necessary to ensure an accurate generation of the fiber modes with the SLM and accurately control various modes in the MMF. In Section 2.1, an algorithm to create phase masks is explained, that allows the generation of arbitrary light fields combining a sequence of SLM pixels, which form a superpixel.

The accuracy of mode excitation is accompanied with the need of an appropriate alignment of the optical setup, which is shown in Figure 1. Hence, a step-by-step instruction is presented in Section 2.2, which shows how an accurate alignment can be achieved.

The light, which exits the MMF, has to be investigated with regard to containing fiber modes. To analyze the received light, one needs access to the complex light field distribution consisting of amplitude and phase information. DH is one promising approach to measure these light properties in a single-shot. In Section 2.3, an algorithm that decomposes the holographically measured light field into the fiber's mode domain is introduced.

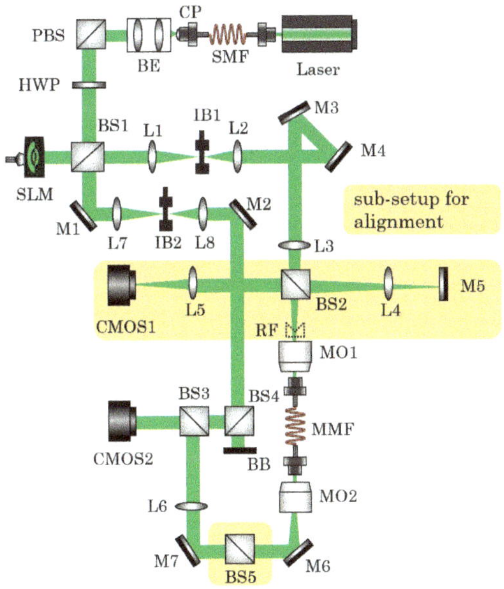

Figure 1. Experimental scheme. The laser beam is coupled to a single mode fiber (SMF), collimated by a collimation package (CP) and widened by a beam expander (BE). A polarizing beam splitter (PBS) and a half-wave plate (HWP) filter the laser beam at a linear polarization, matching the SLM required modulation polarization. Beamsplitter 1(BS1) splits the laser beam into reference and object beam, which is reflected at the SLM plane after being modulated by the SLM pixels. Lens 1 (L1) executes an optical Fourier Transform (FFT) to the modulated laser beam, so a pinhole (IB) filters the first diffraction order in the Fourier plane. L2 executes an optical inverse Fourier Transform (IFFT). L3 and microscope objective 1 (MO1) image the filtered beam onto the fiber facet of the MMF. To yield excitation conditions as precise as possible, the incidence angle of the beam is tuned by two mirrors M3 and M4. The light, that is propagated through the MMF is imaged on camera 2 (CMOS2) with a telescope composed of MO2 and L6. The hologram, which is generated by the interference between the light field that exits the MMF and the reference beam is captured by CMOS2. The reference beam is cleaned up with L7, IB2 and L8. Afterwards, the reference beam is directed onto CMOS2 with M1, M2 and BS4. For proper alignment, the sub-setups in the yellow boxes are used and are explained in detail in Section 2.2. Laser: 532 nm Laser Quantum torus; SMF: THORLABS P3-460B-FC-5; CP: collimator PAF2-A4A; BE: beam expander GBE10-A 10x; HWP: half-wave plate; SLM: HOLOEYE Pluto-2; M1-M6: protected silver mirrors PF10-03-P01; BS1-BS5: 50:50 beam splitter cubes BS013; PBS: polarizing beam splitter cubes PBS251; IB1-2: pinholes D36S; L1-L8: lenses f1 = 250 mm, f2 = 60 mm, f3 = 160 mm, f4 = 80 mm, f5 = 180 mm, f6 = 300 mm, f7 = 60 mm, f8 = 160 mm; MO1: 80x Mitutoyo Plan Apo SL; MO2: Olympus PLN 40x; MMF: THORLABS FG025LJA; BB: beam block; CMOS1 and CMOS2: IDS UI348xLE-M.

2.1. Generating Arbitrary Light Fields Using a Phase-Only SLM

Electromagnetic waves are completely described by amplitude and phase information. With the development of SLMs, the spatial modulation of both the phase and amplitude of light became possible by using a gadget. This possibility has had a huge impact on several optical engineering techniques. It opened the door for coping with optical distortions in flow measurements, investigations of nonlinear phenomena in MMFs or neural networks by shaping the light in a proper way, which is called *adaptive wavefront shaping* (AWS) [15–18]. Within the presented work, AWS is used for an accurate mode excitation by shaping the MMF input wavefield with an SLM. For an ideal mode excitation, the light field projected onto the input facet should correspond to the numerical field distribution as accurately as possible. However, there are no modulators manufactured that are directly intended to shape light's amplitude and phase at the same time. Several techniques have been published that show how to shape a laser beam in amplitude and phase using either an amplitude-only or a phase-only modulator. In [19], a technique is presented, where neighboring modulator pixels form a superpixel to generate a desired complex pointer. In that work, they use a nematic liquid crystal SLM (NLCSLM) to implement a method to control light's amplitude and phase independently. In most NLCSLMs, the phase modulation occurs as a polarization modulation, too. This makes polarization-dependent communication through an MMF difficult. In [20], an amplitude-only modulator based on digital micromirrors is used to achieve a light field modulation. However, a micromirror-based application requires more pixel than a liquid crystal on Silicon (LCoS) SLM to achieve the same degree of freedom in modulation. In this paper, a superpixel technique comparable to the procedure shown in [21,22] using a phase-only LCoS SLM is presented.

In this work, it is assumed that a desired complex field distribution is given as a complex 2D data map that has to be applied on an incident Gaussian laser beam that illuminates the modulator's surface. In the following section, it is explained how the pixelated 2D data map has to be transferred into a 2D superpixel map that can be displayed at the modulator surface to generate a desired complex wavefront.

It is understood that the spatially separated modulation of light of two neighbored SLM pixels is interfering with each other. Depending on the grayscale of the SLM pixels and, hence, the phase shift between the light fractions, the degree of constructive and destructive interference is controlled. Additionally, the resulting phase level is given by the average phase value of the neighbored SLM pixels. Since the SLM plane is imaged on the input fiber facet, one would actually map the individual SLM pixel on the fiber facet. To only map the intensity modulated light, a spatial low-pass filtering is applied [20]. The reflected light is optically Fourier transformed by lens L1. The diffraction orders are spatially separated. Here, the first diffraction order is required and a pinhole (IB) is used to perform low pass filtering (Figure 2).

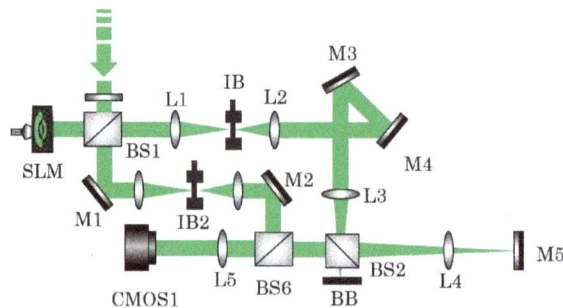

Figure 2. Optical setup for investigation of the superpixel method. This is a sub-setup from the setup shown in Figure 1. For this procedure, BS6 is used to generate a hologram on CMOS1.

The notation of a physical plane wave as a single complex pointer and, furthermore, arbitrary wavefronts being represented by a spatial superposition of several complex pointers are fundamental techniques concerning high frequency engineering. Since visible light is a high frequency electromagnetic wave, it can be described as a complex pointer.

Complex pointers having an arbitrary phase and amplitude can always be represented by a superposition of two other pointers being different from each other, as seen in Figure 3a. Since a superpixel map that can be displayed on a phase-only SLM is required, two complex pointers, whose amplitudes are equal to 1, need to be determined. In superposition, they need to represent the desired data map entry (see Figure 3b). Since the phase distribution of LP modes is binary ($\pi/2$ or $-\pi/2$) [23], the implementation becomes significantly easier. The desired pointer $C(j\omega)$ with the amplitude C_0 from Figure 3b is located either on the positive or negative imaginary axes and can be described as the superposition of two symmetrically displaced unit pointers ($A(j\omega)$ and $B(j\omega)$), whose phase levels Φ_A and Φ_B have to be determined as

$$\Phi_A = arcsin\left(\frac{C_0}{2}\right), \text{ and } \Phi_B = \pi - \Phi_A \qquad (1)$$

The calculated phase levels have to be arranged in a superpixel as shown in Figure 3c.

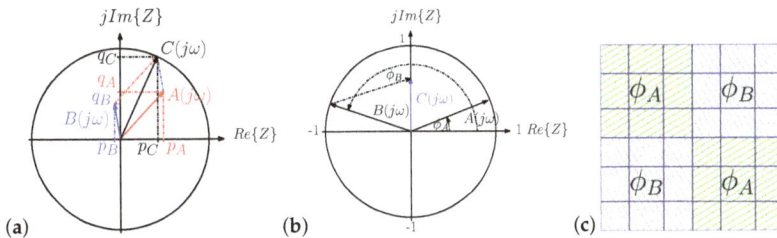

Figure 3. Superposition of complex pointers. (**a**) Generation of an arbitrary complex pointer by superimposing two different complex pointers. (**b**) Generation of a mode data entry with two symmetrically displaced unit pointers for the superpixel map being displayed on a phase-only SLM. (**c**) Resulting superpixel that represents the desired complex pointer $C(j\omega)$. 3 × 3 SLM-Pixel form one subpixel.

The presented principle was applied on a phase-only LCoS SLM from the manufacturer HOLOEYE (type PLUTO-2). In order to image the modulator plane on a 25 µm sized core of a MMF with the optical setup from Figure 2, 935 × 935 pixels from the SLM are used. The required amount of SLM pixels was determined from the demagnification introduced by the optical system of factor 270. The superpixel method was tested using a LP$_{02}$ 'donut' mode as a target field. The measured amplitude and phase distribution are shown in Figure 4a,b. The measured amplitude and phase profiles shown in Figure 4c,d prove that both amplitude and phase gradients can be modulated with the presented technique. The generated field was measured with DH, which is explained in Section 2.3.

One approach to determine the modulation quality of the presented technique is to calculate the fidelity of the measured complex light field $\vec{E}_{\text{Obtained}}$ compared to the numerical distribution $\vec{E}_{\text{Target}}$ [20]. The fidelity is given by

$$Fidelity = F = |\vec{E}_{\text{Obtained}} \vec{E}^*_{\text{Target}}|^2. \qquad (2)$$

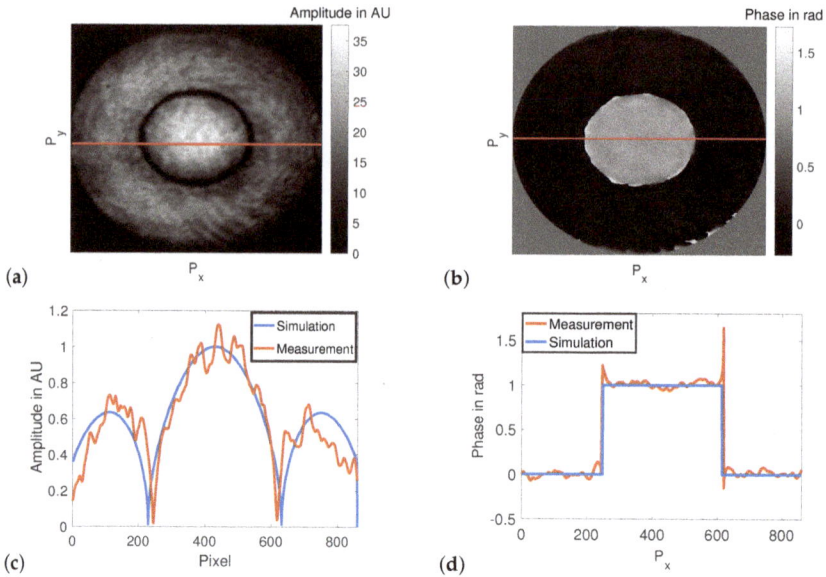

Figure 4. Experimental results from a superpixel-based modulation of a light field. The measured light field is compared with the target LP_{02} mode distribution. (**a**) Amplitude distribution of the measured light field. The red line indicates the analyzed profile route shown in c. (**b**) Phase distribution of the measured light field. The red line indicates the analyzed profile route shown in d. (**c**) Amplitude profile of the measured light field compared with the target profile. (**d**) Phase profile of the measured light field compared with the target profile.

The theoretical maximum of $F_{theoretical} = 0.9943$ was derived by developing a simulation of the presented technique. The computed superpixel pattern is superimposed with a diffraction grating and allocated as the phase distribution of an LP_{01} mode. After performing an FFT, the desired diffraction order is filtered in the Fourier plane with a digital pinhole. Afterwards, an IFFT is performed on the filtered signal. The result is the simulated outcome of the presented technique. Deviations in the simulated result can be explained, on one hand, with both finite aperture and resolution and on the other hand with phase-quantization of the system. In this experiment, a fidelity of 0.9509 is achieved, which corresponds to 95.6% of $F_{theoretical}$. We achieved similar results with modes of higher order. In particular, imperfections of optical components cause deviations from the simulated result. On one hand, LCoS SLMs are known for their phase flickering property [24]. On the other hand, as mentioned above, the first diffraction order is spatially filtered in the Fourier plane behind L1. Undesired polarizations of light and unmodulated light caused by a modulation efficiency < 1 of the SLM lead to unwanted background light and errors, which are located in the 0th diffraction order in the Fourier plane. Because of that, the superpixel distribution is superimposed with a diffraction grating that spatially separates the actual modulated light field into the 1st diffraction order from the 0th order in the Fourier plane. As a consequence, the unwanted light parts can be blocked and the modulated light can pass the pinhole. However, it is necessary to find an optimum trade-off between the superpixel size and the grating constant. The smallest possible binary vertical grating was chosen. The superpixels need to be small enough to achieve a high-performance-modulation, but big enough to not interfere with the grating as shown in Figure 5a. An optimum at 1 Superpixel $= 6 \times 6$ SLM pixels could be observed as shown in Figure 5b.

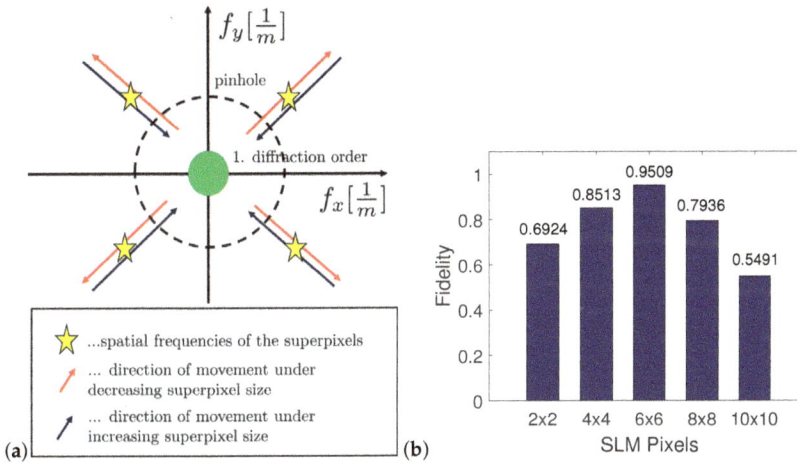

Figure 5. Investigation of superpixel influences on spatial frequencies and resulting fidelity. (**a**) Scheme to illustrate the dependency between spatial frequency movement and superpixel size. (**b**) Measured fidelity of generating a LP_{02} mode by varying superpixel size.

2.2. Specific Alignment Techniques

In this section, a step-by-step instruction that describes how one can image the SLM plane to the input MMF facet plane properly, considering that the planes are aligned in parallel, is presented. The introduced alignment procedure was inspired by the work from [25].

1. Spatial filtering of the modulated light field

    In Section 2.1, it is explained that the derived superpixel distribution needs to be superimposed with a diffraction grating to filter the modulated signal that is located in the 1st diffraction order secluded from the center of the Fourier plane. It is necessary that the filtered signal (first diffraction order) is located on the optical axis to avoid unwanted optical aberrations, which can be achieved by tilting the SLM plane. After conduction, the passed light from IB in Figure 2 propagates through the center of L2.

2. Guiding the laser beam through microscope objective 1 (MO1)

    The utilized subsetup for aligning the laser beam through MO1 is shown in Figure 6. After the IFFT executed by L2, the beam is guided along a desired hole line on the optical bench using two mirrors M3 and M4 and two pinholes (HIB1 and HIB2). Now, L3 can be positioned in the beam path, which should not disturb the beam direction. Finally, the microscope objective can be positioned. In general, microscope objectives magnify small alignment deviations and therefore it is advisable to mount MO1 on a stage with μm precision. Eventually, the beam should propagate through HIB1 and HIB2 after every component was set in the path.

    Furthermore, it is mandatory for a proper mode excitation, that the IFFT with L2 and the telescope involving L3 and MO1 is performing correctly. This depends on the correct distances between the optical components and can be verified by modulating an image with the SLM. In this case, an axicon was used as a target test field. After propagation through the setup, the image behind MO1 can be monitored with a screen. On the screen, the phase boundaries should completely disappear.

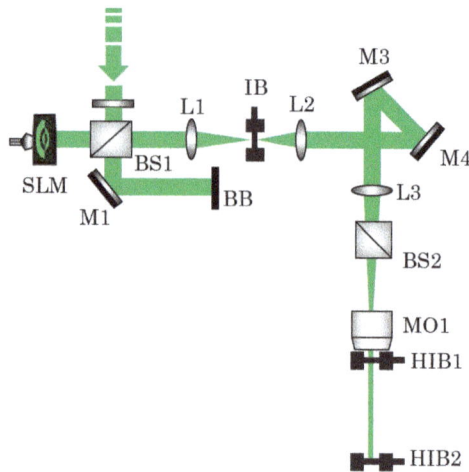

Figure 6. Optical setup to align the microscope objective. This is a sub-setup from the setup shown in Figure 1.

3. Parallel imaging of the SLM plane to MO1

For a proper mode excitation, the SLM plane has to be imaged onto to the input fiber facet. Deviating incident angles could lead to inappropriate mode excitation and should be neglected. Therefore, a non-invasive method to verify a parallel imaging of the mentioned planes was developed. The utilized sub-setup is shown in Figure 7a. The beam propagating through MO1 is split into a second path by BS2, propagates through L4 and is reflected by an auxiliary mirror M5, so that the reflected light is imaged on CMOS1 by another telescope involving L4 and L5. At first, the correct distances between the optical components should be ensured to provide a sharp image of the SLM plane on CMOS1. This was verified by modulating an axicon with the SLM once more. The captured intensity image of the axicon under proper conditions is shown in Figure 7b, where the boundary lines are of minimum thickness. Afterwards, by displaying the sheer diffraction grating on the SLM, the pure modulator plane is imaged on CMOS1 (see Figure 7c). Now, a retro reflector without beam displacement has to be placed in front of MO1 as shown in Figure 7a. Perpendicular alignment of the optical components causes concentric interference rings on CMOS1, as shown in Figure 7d.

4. Alignment of the MMF

With the use of the auxiliary sub-setup introduced in step 3, the MMF can be aligned and a parallel imaging of the SLM plane to the input fiber facet plane is ensured. In the first instance, the MMF has to be set in its place of destination, as shown in Figure 8a. In the next step, light has to be coupled into the fiber at the actual output by using the reference beam. For the alignment, M2, BS5 and M6 can be tilted as desired. The light leaving the fiber input has to be captured with MO1 without moving MO1. This means that only the MMF can be aligned, so that light propagates through MO1 and is captured with CMOS1. In order to find the proper distance between MMF and MO1, the reference beam has to be blocked and an axicon is displayed once more. The back reflecting light from the input fiber facet is now imaged on CMOS1. As introduced in step 3, the MMF has to be moved until the boundary lines of the axicon are at a minimum thickness, as shown in Figure 8b. Now, the input fiber facet is located in the focal plane of MO1. By coupling the reference beam into the actual fiber output once more, a sharp image of the fiber input facet on CMOS1 can be seen, as shown in Figure 8c. Eventually, if the sheer diffraction grating is displayed on the SLM and the back reflected light from the input fiber facet interferes with the reflection from M5, concentric interference rings appear on CMOS1, as shown in Figure 8d.

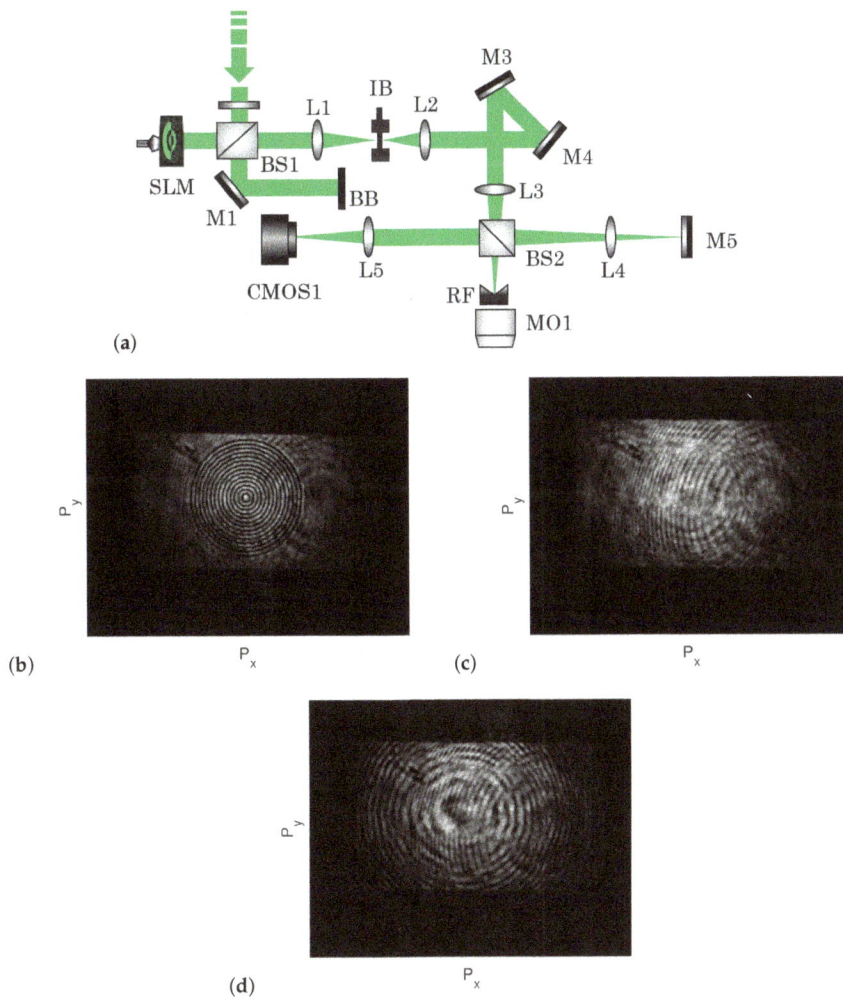

(a)

(b)

(c)

(d)

Figure 7. An auxiliary sub-setup helps to image the SLM plane to the input fiber facet plane. At first, the auxiliary path with L4 and M5 needs to be set properly: (**a**) sub-setup from Figure 1; (**b**) sharp image from an axicon reflected by the auxiliary mirror M5 and captured with CMOS1; (**c**) sharp image of the modulator plane reflected from M5 and captured with CMOS1. The SLM displays the sheer diffraction grating; (**d**) after the retro reflector is set, concentric interference rings appear on CMOS1.

(a)

(b)

(c)

(d)

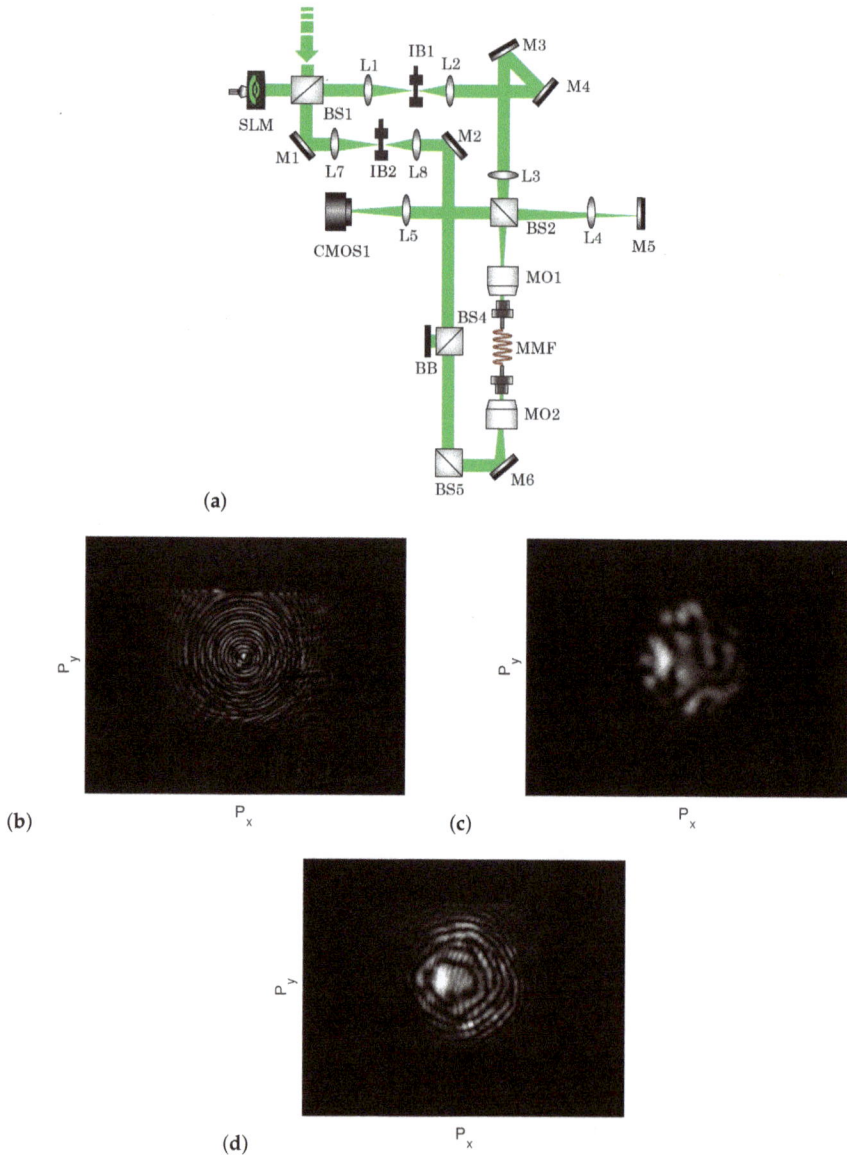

Figure 8. For the alignment of the MMF, the reference beam is coupled to the actual output of the MMF and the image from the input fiber facet on CMOS1 is analyzed: (**a**) sub-setup from Figure 1; (**b**) sharp image from an axicon reflected by the input fiber facet and captured with CMOS1; (**c**) sharp image of the input fiber facet and captured with CMOS1; (**d**) after the MMF is aligned properly, concentric interference rings appear on CMOS1 if the back reflected light from the input fiber facet interferes with the reflected light from M5. The sheer diffraction grating is displayed on the SLM.

2.3. Holographic Decomposition of Light into the Fiber's Mode Domain

DH is one possible approach to extract amplitude and phase from an electromagnetic wave using a CMOS digital camera [26]. Applying this technique to the presented experimental setup, the fiber output light field can be measured and analyzed. Therefore, the MMF output signal is superimposed with the reference beam on CMOS2, as shown in Figure 1. The resulting hologram is captured with the camera and numerically reconstructed with the angular spectrum method [27].

In order to receive a quantitative phase measurement, the raw phase data from the measurement has to be relieved from several errors. These unwanted systematic parts of the measured phase data are due to wrapping errors, optical aberrations, phase offsets, MMF surface irregularities or dust on optical components. One approach to determine these error terms is to capture a reference hologram to identify a reference phase distribution, which can be set against the experimental results. In Section 3, it is explained how this is conducted during the TM acquisition.

For acquiring a fiber's TM, the holographically measured complex light field has to be decomposed into the fiber's mode domain. Since Maxwell's equations are linear differential equations, the light guidance of an arbitrary input field $\underline{E}_{input}$ in an MMF can be modeled as a linear sequence approach, where the orthogonal base functions are the complex spatial distributions of the supported fiber modes $\underline{E}_n$ [28]. The proportion of every mode to $\underline{E}_{input}$ is determined by their individual weight a_n:

$$\underline{E}_{input} = \sum_{n=1}^{N} a_n \underline{E}_n = a_1 \underline{E}_1 + a_2 \underline{E}_2 + ... + a_N \underline{E}_N. \tag{3}$$

Likewise, the measured field at the fiber output $\underline{E}_{obtained}$ consisting of the measured amplitude $|A|$ and phase ϕ distributions,

$$\underline{E}_{obtained} = |A| \cdot e^{i\phi}, \tag{4}$$

needs to be analyzed considering the supported fiber modes $\underline{E}_n$ from the same mode domain:

$$\underline{E}_{obtained} = \sum_{n=1}^{N} c_n \underline{E}_n = c_1 \underline{E}_1 + c_2 \underline{E}_2 + ... + c_N \underline{E}_N. \tag{5}$$

The coefficients c_n determine the weight of the individual modes in the measured light field. In order to identify the coefficient of every mode in a measured light field, a linear system of equations is derived by multiplying the complex conjugate base functions $\underline{E}_1^*, \underline{E}_2^*, ...,$ and integrate over the whole space, named R. The integration is approximated by a two-dimensional sum, thus the camera captures a pixelated image:

$$\sum_R \underline{E}_{obtained}\underline{E}_1^* = c_1 \sum_R \underline{E}_1\underline{E}_1^* + c_2 \sum_R \underline{E}_2\underline{E}_1^* + ... + c_N \sum_R \underline{E}_N\underline{E}_1^*$$
$$\sum_R \underline{E}_{obtained}\underline{E}_2^* = c_1 \sum_R \underline{E}_1\underline{E}_2^* + c_2 \sum_R \underline{E}_2\underline{E}_2^* + ... + c_N \sum_R \underline{E}_N\underline{E}_2^*$$
$$... = ... \tag{6}$$
$$\sum_R \underline{E}_{obtained}\underline{E}_N^* = c_1 \sum_R \underline{E}_1\underline{E}_N^* + c_2 \sum_R \underline{E}_2\underline{E}_N^* + ... + c_N \sum_R \underline{E}_N\underline{E}_N^*.$$

The solution of this system becomes much easier due to the fact that the base functions are orthogonal to each other hence:

$$\sum_R \underline{E}_n\underline{E}_m^* = \begin{cases} 0 & \text{for } n \neq m \\ K_n & \text{for } n = m \end{cases} \tag{7}$$

Therefore, the right hand side of every row n of the system (6) disappears, except the item $c_n K_n$ and so, the coefficients can be calculated by

$$c_n = \frac{1}{K_n} \sum_R \underline{E}_{obtained}\underline{E}_n^*. \tag{8}$$

If Equation (8) is computed for every mode from the N-dimensional mode domain, the measured distribution $\underline{E}_{\text{obtained}}$ is decomposed into the fiber's mode domain by single-shot.

3. Acquisition of the Fiber's Transmission Matrix

In the previous section, the experimental principles required for the direct TM acquisition were introduced. It is required to excite every mode of the supported fiber modes sequentially and directly without the need to use a conversion matrix as done in [11] or [12]. Selective mode excitations are especially beneficial for prospective information-technical applications, however the experimental work appears more challenging. The actual procedure of the TM acquisition is based on the work from [12]. To verify the developed technique with the experimental setup from Figure 1, a TM acquisition at an approximately 1 cm long piece of step-index MMF is presented. MMFs at this size have a weakly developed mode scrambling property [11]; thus, an acquired TM corresponds strongly to a diagonal matrix. This characteristic is used to provide a proof-of-principle of the presented system.

After the complex spatial mode functions from every supported fiber mode have been calculated and translated into a superpixel distribution, a reference phase distribution for the holographic-based decomposition method introduced in Section 2.3 is obtained. In order to measure systematic aberrations, it is necessary to let the object beam, which should be as unmodulated as possible, propagate through the optical setup to measure all the aberrations that act on the object beam. Since the errors, which have to be compensated, are located behind the MMF output, an unmodulated object beam at the fiber output needs to be generated. Due to the weakly developed scrambling at MMFs of approximately 1 cm length, the LP_{01} mode is excited at the fiber input. The light that propagated through the MMF appears with a sufficiently homogeneous phase distribution at the MMF output and can be obtained holographically at CMOS2. The reconstructed amplitude and phase distributions are shown in Figure 9. Now, the correction can be conducted by calculating $\phi_{\text{corrected}} = \phi_{\text{measured}} - \phi_{\text{Ref}}$.

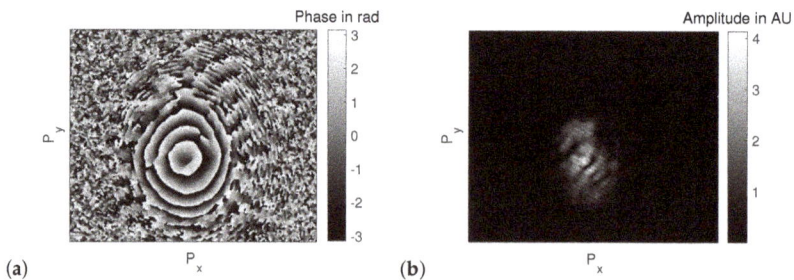

Figure 9. Reconstruction of the reference hologram captured with CMOS2 by exciting an LP_{01} mode at the fiber input: (**a**) reconstructed phase distribution; (**b**) reconstructed amplitude distribution.

With access to a quantitative phase measurement, the TM acquisition can be executed. The fiber under study supports $N = 110$ fiber modes considering [28], where N is calculated by:

$$N \approx \frac{1}{2} \cdot \left(a \cdot \frac{2\pi}{\lambda_0} \cdot \text{NA} \right)^2 , \tag{9}$$

with a being the MMF core radius. The fiber specifications are listed in Table 1.

Table 1. Specifications of the used MMF.

Specification	Value
NA	0.1 ± 0.015
a	(25 ± 3) μm
λ_0	532 nm

Hence, the expected TM is of dimensions 110×110, which means that for each polarization direction a 55×55 TM can be acquired. Now, every mode contained in the fiber's mode domain has to be excited sequentially at the fiber input. The propagated light fields are superimposed with the reference beam on CMOS2, where holograms are captured. The holograms are reconstructed numerically with the angular spectrum method. Afterwards, the decomposition has to be executed for every shot. The resulting coefficients from the *i*-th mode of the mode domain are stored in the *i*-th row of the TM. Within the introduced experiment, the TM shown in Figure 10 was measured. Both the input and output vectors with the determined coefficients are arranged following the mode group manner presented in [7]. One result was that every LP_{lm} mode belongs to a mode group given by $2m + l - 1$. Modes inside a group mix heavily with each other.

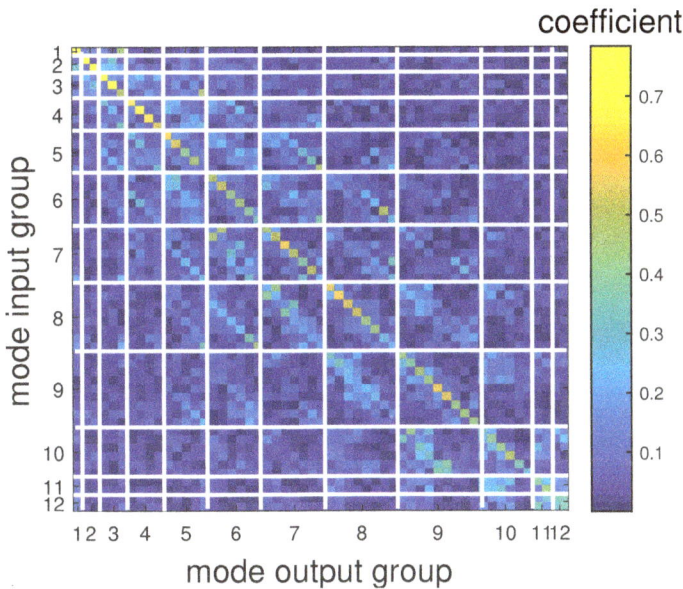

Figure 10. Measured TM of one linear polarization direction of a MMF. During the experiment, constant laboratory conditions were guaranteed.

4. Discussion

The introduced superpixel method was implemented in the presented optical setup. This technique enables arbitrary complex modulation with one phase-only SLM. Therefore, a high-fidelity mode generation is possible. Hence, the supported fiber modes can be directly excited at the fiber input. A theoretical maximum of the presented technique of $F_{theoretical} = 0.9943$ was derived. During the experiment, a value of $F = 0.9509$ could be achieved, which corresponds to 95.6% of $F_{theoretical}$. The main part of occurring deviations in the simulation can be explained with the overlaying diffraction grating on the superpixel pattern to reduce the influences of undesired polarizations of light or unmodulated light. This leads to a trade-off between the size of the superpixel and the grating. The selective mode excitation is followed by a challenging alignment procedure, which could be managed by developing specific adjustment methods. A DH-based decomposition technique that is used to analyze the propagated light with respect to containing fiber modes was introduced. Hence, one row of the acquired TM can be obtained in a single-shot and reduces the measurement effort by a factor of 4 in relation to phase-shifting-based acquisition methods or factor N in relation to correlation filter methods. Presently, the bottle neck regarding the TM acquisition time is the SLM's frame rate (best performance at approximately 20 Hz). Eventually, a TM measurement of an approximately 1 cm long piece of MMF was conducted. At these propagation distances, mode coupling effects are weakly developed, so that a TM resembles a diagonal matrix so long as laboratory conditions are constantly complied. The major power of the resulting TM, which is shown in Figure 10, is clearly distributed over the diagonal elements. Hence, a proof of the presented principle could be accomplished. However, TM entries off the diagonal have a value significantly greater than background noise and need to be discussed. By using a piece of MMF, which has a length of approximately 1 cm, it was intended to reduce the scattering property of the investigated sample. Yet, the utilized laser with $\lambda_0 = 532$ nm propagates with approximately 20,000 waves through the waveguide, which is still enough to come upon manufacturing imperfections. In Table 1, the fiber properties with the given tolerances are listed. Both the core diameter and the NA have a tolerance of more than 10%—these are imperfect conditions. In [11], the authors obtained similar results: the simulated phase distribution after propagation through an approximately 10 cm piece of step-index MMF did not fit the experimental results until they superimposed the ideal refractive index profile with Zernike polynomials. As a consequence, the person who works with MMFs should address manufacturing imperfections. The presented results open the door to fast TM acquisition procedures of highly moded MMFs using MDM. In such procedures, the amplitude and phase relations are determined to understand the mode scrambling processes inside the MMF affecting the TM.

5. Conclusions

An optical setup in order to excite N-supported fiber modes of a step-index MMF with a phase-only SLM using superpixels was investigated. With a novel decomposition method based on digital holography, the TM of a MMF is measured with only N measurements. This paves the way for prospective rapid TM measurements. The TM of MMF provides important channel information for the implementation of Mode Division Multiplexing, which is based on Multiple Input Multiple Output (MIMO). As a perspective, MIMO can be used to increase the secure goodput of the data transfer based on physical layer security.

Author Contributions: J.W.C. and N.K. contributed to the idea of using digital holographic methods to measure the transmission matrix of multimode fibers. S.R., N.K. and H.R. designed, built, programmed and characterized the setup. S.R. wrote the article. J.W.C. supervised the whole research work. N.K., H.R. and J.W.C. revised the article.

Funding: This research was funded by Deutsche Forschungsgemeinschaft (DFG) grant number (CZ 55/42-1).

Conflicts of Interest: The authors declare no conflict of interest.

References

1. Papadopoulos, I.N.; Farahi, S.; Moser, C.; Psaltis, D. Focusing and scanning light through a multimode optical fiber using digital phase conjugation. *Opt. Express* **2012**, *20*, 10583–10590. [CrossRef] [PubMed]
2. Haufe, D.; Koukourakis, N.; Büttner, L.; Czarske, J. Transmission of multiple signals through an optical fiber using wavefront shaping. *J. Vis. Exp.* **2017**, *121*, e55407. [CrossRef] [PubMed]
3. Czarske, J.W.; Haufe, D.; Koukourakis, N.; Büttner, L. Transmission of independent signals through a multimode fiber using digital optical phase conjugation. *Opt. Express* **2016**, *24*, 15128–15136. [CrossRef] [PubMed]
4. Florentin, R.; Karmene, V.; Benoist, J.; Desfarges-Berthelemot, A.; Pagnoux, D.; Barthélémy, A.; Huignard, J.P. Shaping the light amplified in a multimode fiber. *Light Sci. Appl.* **2017**, *6*, e16208. [CrossRef] [PubMed]
5. Berdagué, S.; Facq, P. Mode division multiplexing in optical fibers. *Appl. Opt.* **1982**, *21*, 1950–1955. [CrossRef] [PubMed]
6. Ryf, R.; Fontaine, N.K.; Chen, H.; Guan, B.; Huang, B.; Esmaeelpour, M.; Gnauck, A.H.; Randel, S.; Yoo, S.J.B.; Koonen, A.M.J.; et al. Mode-multiplexed transmission over conventional graded-index multimode fibers. *Opt. Express* **2015**, *23*, 235–246. [CrossRef] [PubMed]
7. Carpenter, J.; Thomsen, B.C.; Wilkinson, T.D. Degenerate Mode-Group Division Multiplexing. *J. Lightw. Technol.* **2012**, *30*, 3946–3952. [CrossRef]

8. Richardson, D.J.; Fini, J.M.; Nelson, L.E. Space-division multiplexing in optical fibres. *Nat. Photonics* **2013**, 7, 354–362. [CrossRef]
9. Jorswieck, E.; Wolf, A.; Gerbracht, S. Secrecy on the Physical Layer in Wireless Networks. In *Trends in Telecommunications Technologies*; INTECH: Boston, MA, USA, 2010; pp. 413–435.
10. Popoff, S.M.; Lerosey, G.; Carminati, R.; Fink, M.; Boccara, A.C.; Gigan, S. Measuring the Transmission Matrix in Optics: An Approach to the Study and Control of Light Propagation in Disordered Media. *Phys. Rev. Lett.* **2010**, *104*, 100601–100605. [CrossRef] [PubMed]
11. Plöschner, M.; Tyc, T.; Čižmár, T. Seeing through chaos in multimode fibres. *Nat. Photonics* **2015**, *9*, 529–535. [CrossRef]
12. Gu, R.Y.; Mahalati, R.N.; Kahn, J.M. Design of flexible multi-mode fiber endoscope. *Opt. Express* **2015**, *23*, 26905–26918. [CrossRef] [PubMed]
13. Flamm, D.; Schulze, C.; Naidoo, D.; Schröter, S.; Forbes, A.; Duparré, M. All-Digital Holographic Tool for Mode Excitation and Analysis in Optical Fibers. *J. Lightw. Technol.* **2013**, *31*, 1023–1032. [CrossRef]
14. N'Gom, M.; Norris, T.B.; Michielssen, E.; Nadakuditi, R.R. Mode control in a multimode fiber through acquiring its transmission matrix from a reference-less optical system. *Opt. Lett.* **2018**, *43*, 419–422. [CrossRef]
15. Koukourakis, N.; Fregin, B.; König, J.; Büttner, L.; Czarske, J.W. Wavefront shaping for imaging-based flow velocity measurements through distortions using a Fresnel guide star. *Opt. Express* **2016**, *24*, 22074–22087. [CrossRef] [PubMed]
16. Kuschmierz, R.; Scharf, E.; Koukourakis, N.; Czarske, J.W. Self-calibration of lensless holographic endoscope using programmable guide stars. *Opt. Lett.* **2018**, *43*, 2997–3000. [CrossRef] [PubMed]
17. Tzang, O.; Caravaca-Aguirre, A.M.; Wagner, K.; Piestun, R. Adaptive wavefront shaping for controlling nonlinear multimode interactions in optical fibres. *Nat. Photonics* **2018**, *12*, 368–374. [CrossRef]
18. Schmieder, F.; Klapper, S.D.; Koukourakis, N.; Busskamp, V.; Czarske, J.W. Optogenetic Stimulation of Human Neural Networks Using Fast Ferroelectric Spatial Light Modulator—Based Holographic Illumination. *Appl. Sci.* **2018**, *8*, 1180. [CrossRef]
19. Van Putten, E.G.; Vellekoop, I.M.; Mosk, A.P. Spatial amplitude and phase modulation using commercial twisted nematic LCDs. *Appl. Opt.* **2008**, *47*, 2076–2081. [CrossRef]
20. Goorden, S.A.; Bertolotti, J.; Mosk, A.P. Superpixel-based spatial amplitude and phase modulation using a digital micromirror device. *Opt. Express* **2014**, *22*, 17999–18009. [CrossRef]
21. Forbes, A.; Dudley, A.; McLaren, M. Creation and detection of optical modes with spatial light modulators. *Adv. Opt. Photonics* **2016**, *8*, 200–227. [CrossRef]
22. Chu, D.C.; Goodman, J.W. Spectrum Shaping with Parity Sequences. *Appl. Opt.* **1972**, *11*, 1716–1724. [CrossRef] [PubMed]
23. Brüning, R.; Flamm, D.; Ngcobo, S.S.; Forbes, A.; Duparré M. Rapid measurement of the fiber's transmission matrix. *Proc. SPIE* **2015**, *9389*, 93890N .
24. García-Márquez, J.; López, V.; González-Vega, A.; Noé, E. minimization in an LCoS spatial light modulator. *Opt. Express* **2012**, *20*, 8431–8441. [CrossRef] [PubMed]
25. Jang, M.; Ruan, H.; Zhou, H.; Judkewitz, B.; Yang, C. Method for auto-alignment of digital optical phase conjugation systems based on digital propagation. *Opt. Express* **2014**, *22*, 14054–14071. [CrossRef] [PubMed]
26. Schnars, U.; Jüptner, W. Direct recording of holograms by a CCD target and numerical reconstruction. *Appl. Opt.* **1994**, *33*, 179–181. [CrossRef] [PubMed]
27. Koukourakis, N.; Abdelwahab, T.; Li, M.Y.; Höpfner, H.; Lai, Y.W.; Darakis, E.; Brenner, C.; Gerhardt, N.C.; Hofmann, M.R. Photorefractive two-wave mixing for image amplification in digital holography. *Opt. Express* **2011**, *19*, 22004–22023. [CrossRef] [PubMed]
28. Gloge, D. Weakly Guiding Fibers. *Appl. Opt.* **1971**, *10*, 2252–2258. [CrossRef] [PubMed]

*applied
sciences*

MDPI

Review

Microparticle Manipulation and Imaging through a Self-Calibrated Liquid Crystal on Silicon Display

Haolin Zhang [1,*], Angel Lizana [1], Albert Van Eeckhout [1], Alex Turpin [2,3], Claudio Ramirez [1,4], Claudio Iemmi [5] and Juan Campos [1]

[1] Departamento de Física, Universitat Autònoma de Barcelona, 08193 Bellaterra, Spain;
    angel.lizana@uab.cat (A.L.); albert.vaneeckhout@uab.cat (A.V.E.);
    claudio.ramirez@ccadet.unam.mx (C.R.); juan.campos@uab.cat (J.C.)
[2] Leibniz Institute of Photonic Technology, Albert-Einstein-Straße 9, 07745 Jena, Germany
[3] School of Physics and Astronomy, Kelvin Building, University of Glasgow, Glasgow G12 8QQ, UK
[4] Instituto de Ciencias Aplicadas y Tecnología (ICAT), Universidad Nacional Autónoma de México,
    04510 Mexico City, Mexico
[5] Departamento de Física, Universidad de Buenos Aires, Consejo Nacional de Investigaciones Científicas y
    Técnicas, 1428 Buenos Aires, Argentina; iemmi@df.uba.ar
* Correspondence: haolin.zhang@uab.cat; Tel.: +34-633-582-601

Received: 24 October 2018; Accepted: 16 November 2018; Published: 20 November 2018

Abstract: We present in this paper a revision of three different methods we conceived in the framework of liquid crystal on silicon (LCoS) display optimization and application. We preliminarily demonstrate an LCoS self-calibration technique, from which we can perform a complete LCoS characterization. In particular, two important characteristics of LCoS displays are retrieved by using self-addressed digital holograms. On the one hand, we determine its phase-voltage curve by using the interference pattern generated by a digital two-sectorial split-lens configuration. On the other hand, the LCoS surface profile is also determined by using a self-addressed dynamic micro-lens array pattern. Second, the implementation of microparticle manipulation through optical traps created by an LCoS display is demonstrated. Finally, an LCoS display based inline (IL) holographic imaging system is described. By using the LCoS display to implement a double-sideband filter configuration, this inline architecture demonstrates the advantage of obtaining dynamic holographic imaging of microparticles independently of their spatial positions by avoiding the non-desired conjugate images.

Keywords: Liquid Crystal on Silicon display; phase modulation; optical manipulation; calibration; holography; diffractive optics

1. Introduction

The interest of using Liquid Crystal on Silicon (LCoS) displays to implement wavefront modulation has been widely discussed in literature [1–4]. LCoS displays are reflective Liquid Crystal Displays (LCDs) that can be customized to operate as spatial light modulators (SLMs). Into the LCoS architecture, LC molecules are evenly distributed on a thin layer with a pixelated aluminum array connected beneath. Underneath the aluminum pixel array, an electronic circuity is integrated into the silicon chip, which allows controlling the voltages addressed to any pixel in the display [3]. Once the voltage is driven to the LC molecules in each pixel, they tend to align with the applied electric field. Importantly, LC molecules orientation varies with the applied voltage, and thus, different phase retardation is introduced as a function of the voltage. When the system is optimized to work into the phase-only regime (by controlling the polarization of light illuminating the display [5]), the wavefront phase distribution can be spatially modified by addressing the proper voltage-inspired phase retardation distribution.

Note that LCoS displays can be configured to both operating in the amplitude-only or the phase-only regimes. Into the amplitude modulation scheme, output intensity distribution can be customized by properly modifying the polarization spatial distribution of the wavefront, and projecting it over a linear analyzer, being this configuration commonly used for imaging proposes. Conversely, LCoS displays operating in the phase-only regime are commonly used for phase modulation, being of interest in a wide number of applications, such as microstructure fabrication [6], holography [7,8], biomedical applications [9], waveguide optics [10], optical switching [11], microparticle manipulation [12,13], etc. To guarantee an optimal working performance in all those applications, LCoS must be previously calibrated and optimized. In particular, the phase-voltage relation and the LCoS surface homogeneity (flatness) are two important characteristics to be calibrated as they directly determine the performance of the generated phase distribution [14–16]. In this paper, we review a novel method able to provide both characteristics in a compact and feasible optical arrangement. The method is based on the use of diffractive optical elements (DOEs) self-addressed to the LCoS display to be optimized. Here, the term "self-addressing" means that the used DOEs are addressed to the same LCoS display to be calibrated, this leading to an auto-calibrating procedure. In particular, two DOEs are self-addressed: (1) a split-lens configuration [17,18] to fulfil the phase-voltage calibration; and (2) a spot-array pattern (Shack–Hartmann, S-H, configuration [19–21]) to realize the surface profile measurement. For the phase-voltage calibration case, a diffractive two split-lens configuration is addressed on the LCoS, which is illuminated by coherent light. This configuration is equivalent to the Young's double slit experiment, so the well-known fringes pattern is realized to a propagated plane. Later, by adding different constant phases (by driving different voltages) to one of the two split-lenses phase distributions, the fringes pattern is shifted in the transversal direction according to the added phase. Under this scenario, the LCoS phase-voltage relation is directly obtained by analyzing the interferometry pattern shift within each driven voltage. Here, we want to note that our interferometry-based technique demonstrates great advantage compared to other existing schemes. For instance, the method is able to take into account the unsatisfied time-fluctuation effect (also refereed as flicker effect [22]) that degrades the performance of LCoS, without requiring extra metrological set-ups (i.e., the system is self-calibrated). For the surface profile calibration case, a highly collimated beam illuminates the LCoS display where an S-H configuration has been implemented. After reflection on the LCoS display, the input light creates a spot array distribution at the micro-lens focal plane. If the input light is highly collimated, light dots deviations from a reference dots-pattern are mainly originated by inhomogeneity at the LCoS screen. At this moment, by calculating the deviation of each individual spot to its theoretical center, the surface profile can be accurately retrieved by using numerical methods.

Once the LCoS display is optimized, we use it for optical applications. Two different applications we recently developed for microparticles imaging [23,24] and manipulation [12,13] are revised in this manuscript. First, we use such elements to realize customized three-dimensional light structures. Due to photophoretic force fields [25] associated to created light structures, we conduct microparticle spatial trapping and manipulation, being useful to investigate physical or biological phenomenon of molecules [26], atoms [27], or even cells [28]. To obtain such light trapping architecture, we propose split-lens inspired configurations addressed to the LCoS display. Second, we propose an inline (IL) optical arrangement to perform dynamic holographic imaging of microparticles (placed to arbitrary spatial positions). In particular, this IL system is based on LCoS device used to generate a double-sideband filter, to be applied at the Fourier plane of the holographic system [29]. Under this scenario, the undesired conjugate image (ghost image) is dynamically eliminated by the double-sideband filter. When compared with other proposals, our optical arrangement presents significant advantages, as the elimination of ghost images into an inline and robust configuration, and without the necessity of time-sequential measurements.

The outline of this review is as follows. In the Materials and Methods section, we preliminarily demonstrate the LCoS display calibration technique, with special focus on the phase-voltage relation calibration and the surface homogeneity determination. We also provide the use of this device to implement an optical architecture to realize particle trapping and manipulation. Finally, this section also presents the description and the usage of the double-sideband filter based inline holographic system, for microparticles holographic imaging. In the Results section, we show the experimental results obtained for the three methods described. We firstly present the LCoS display self-calibration results, for the phase-voltage look-up table and the LCoS screen profile. Afterwards, we show the experimental realization of microparticles trapping and spatial control with this calibrated LCoS display. Finally, by registering the LCoS display into the IL holographic architecture, we demonstrate the real-time holographic imaging of microparticles which are located in different spatial positions. In the Conclusion section, we highlight the important role of the LCoS display in the diverse methods proposed.

2. Materials and Methods

2.1. LCoS Display Self-Calibration

In order to optimally apply any LCoS display into a wavefront modulation system, the characteristic parameters of the LCoS display must be calibrated. A method for determining both the phase-voltage look-up table (Section 2.1.1) and the surface profile (Section 2.1.2) of the LCoS is following described.

2.1.1. Phase-Voltage Calibration

We use an interferometric method, conducted by self-addressing a reconfigurable two-sectorial split-lens scheme on the LCoS display to be calibrated, to determine its phase-voltage response [15].

The two-sectorial split-lens configuration applied here represents the classic Billet lens scheme [30], which leads to an interference pattern equivalent to that of the Young's double slit experiment. The optical scheme of this method is sketched in Figure 1. The Billet lens scheme consists of a lens cut from its center into two sections and then being separated with a certain distance (Figure 1b). The phase distribution of the digitally generated split-lens scheme, to be addressed to the LCoS display, is depicted in Figure 1a. Under this scenario, the coherent light illuminating the two-sectorial split-lens configuration at the LCoS plane generates two corresponding focal points (F1 and F2) at the focal plane (S plane in Figure 1). These two focal points act as new coherent and punctual light sources. Later, these two divergent beams further propagate and finally encounter at plane B, where a fringes-like interferometric pattern is produced (Figure 1c).

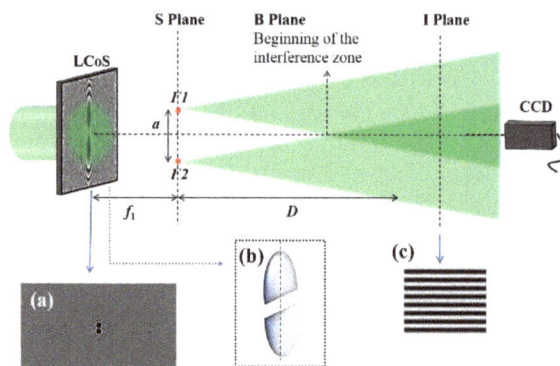

Figure 1. Two-sectorial split-lens based optical scheme: (**a**) Phase distribution of the split-lens scheme on the LCoS display; (**b**) Billet lens scheme; (**c**) Fringe-like interferometric pattern.

Once the interferometry pattern is obtained, the phase-voltage relation can be retrieved. To better explain this method, let us review the mathematical expression describing the two split-lens phase distribution shown in Figure 1a. This is given in Equation (1).

$$U_{N=2}(r,\theta) = U_1 + U_2,$$ (1)

where U_1 and U_2 are the phase distribution for each one of the two lens sectors and they are written as [15],

$$U_1 = \exp[i\frac{\pi}{\lambda f}(r^2 + a_0{}^2 - 2ra_0\cos(\theta_i - \theta_0) + \phi(V))],$$ (2)

$$U_2 = \exp[i\frac{\pi}{\lambda f}(r^2 + a_1{}^2 - 2ra_1\cos(\theta_{ii} - \theta_1))],$$ (3)

where f refers to the focal length of the split lens, r to the radial coordinate in the lens plane and λ represents the wavelength for the light source. In addition, a_0 is the distance from the first split-lens sector center to the LCoS geometric center, while a_1 is the distance from the second split-lens sector center to the same LCoS center. What is more, θ_0 and θ_1 refers to the angular positions of the two split sector centres. Thus, the upper sector angular distribution θ_i is restricted in the range as π to 2π, whereas the lower sector angular distribution θ_{ii} is restricted in the range as 0 to π. Note that these two sectors fully cover the whole angular distribution as 0 to 2π. Finally, note that the phase distribution for the first split sector (Equation (2)) introduces a constant phase term which depends on the voltage, $\phi(V)$.

By setting a term $\phi(V)$ different to zero in Equation (2) (each possible $\phi(V)$ is related to a different gray level), a constant phase is added to one of the split sectors (U_1 in our case). This produces an extra retardance between wavefronts originated by U_1 and U_2, which is transferred into a transversal shift of the fringe pattern in a given axial plane (e.g., I plane). Therefore, by measuring the transversal shifts corresponding to the gray levels, the phase-gray levels relation can be determined, for instance, by using correlation-based calculations [22].

2.1.2. Surface Profile Calibration

To conduct the LCoS display surface calibration, a digital micro-lens array pattern (Shack–Hartmann scheme [19,20]) is self-addressed to the device. The micro-lens array consists of an ordered lattice of micro diffractive convergent lenses integrated at the LCoS plane. Moreover, all these lenses share the same focal length and the distance between any two adjacent lenses is equivalent. In this scenario, by sending the proper S-H phase distribution to the LCoS display, the diffractive micro-lens array is created. Let us assume now that we illuminate the LCoS with a collimated beam perpendicular to the display. In the case of a perfectly flat LCoS display surface (i.e., the screen does not introduce any extra phase distribution), we would obtain a uniformly distributed spot light configuration at the focal plane of the micro-lenses. In other words, each fraction of the input plane wave is focused at the optical axis of its corresponding micro-lens (see Figure 2a). On the contrary, if the LCoS screen presents spatial defects, the plane input wave is spatially distorted, and an extra phase distribution is added to the system, in a situation equivalent to Figure 2b. In such a scenario, corresponding spot lights at the focal plane are not evenly distributed but presents some deviation from the expected centers. Finally, by calculating the deviations of light dots with respect to the expected centers (see Figure 2b), the surface profile can be retrieved by using proper numerical methods.

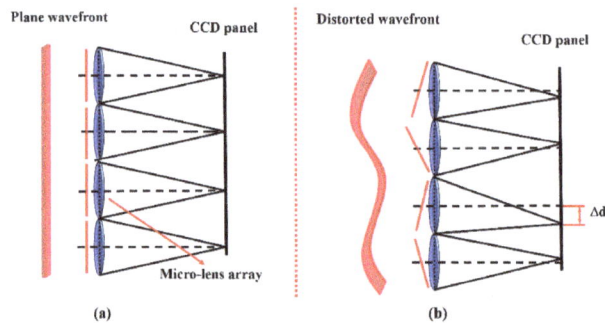

Figure 2. Shack–Hartmann micro-lens array scheme: (**a**) Plane wavefront modulated by the diffractive micro-lens array and the spot light configuration demonstrates a uniform distribution; (**b**) Distorted wavefront modulated by the diffractive micro-lens array and the spot light configuration demonstrates the spatial deviation.

We determine the light spot deviations by using centroid calculations thoroughly described in Ref. [20]. The main idea is to use a reference grid at the charge-coupled device (CCD) plane that indicates the theoretical centers. In particular, this reference grid corresponds to a system free of aberrations (light distribution at the focal plane related to a plane wave). Later, this grid is regarded as the criteria and the actual deviations of light dots observed when illuminating the LCoS, are estimated with respect to this reference grid. By determining such deviation, well-known integration methods [31] are used to retrieve the surface of the LCoS.

Two problems may arise by using this approach. First, we may introduce a reference grid pattern which is shifted from the theoretical grid. This situation leads to a constant shift into the centroid calculation. Nevertheless, this issue is easy to solve as in fact such grid shift can be regarded as a constant value added to the deviation function, and therefore, it only introduces an artificial tilt after using integration methods. Therefore, this situation does not modify the calculated surface profile. Conversely, we may face a second problem. If the size of the grid squares we choose is different from the theoretical one, linear phases are introduced into the centroid calculation and thus, a quadratic function is added to the surface profile after reconstruction during the integration. Therefore, the size of the grid has to be correctly determined as an input parameter to properly retrieve the LCoS surface profile. To solve this situation, we used the grid size calibration method discussed in Ref. [15]. Once the theoretical grid is properly set, experimental light spot deviations are measured with respect to theoretical centers. At this point, we can easily calculate the derivative function (slopes of the actual surface) from shifted light spots and then retrieve the surface profile by simply using the two-dimensional integration. At last, we use cubic splines [31] to interpolate discrete data and obtain a continuous surface profile.

2.2. Microparticle Manipulation through Light Structures Created by a LCoS Display

We also present a method for microparticle spatial manipulation based on an LCoS display [12]. Before particle manipulation, they are firstly optically trapped into a three-dimensional light structure (optical bottle structure). These light structures are created thanks to the split-lens inspired configuration addressed to the LCoS. The physical mechanism that guarantees the particle trapping is here explained. The particle heated by the optical bottle structure experiences an opto-thermal force (photophoretic force [25]) with its force direction opposite to the gravity. Therefore, once the gravity of the particle is compensated by the photophoretic force, the particle is suspended in the air.

To obtain the reliable, easy-implemented light trapping architecture, we send to the LCoS display a digital diffractive *N*-sector based spilt-lens pattern [30,32]. This *N*-sector split-lens scheme shares the same principle as the Billet lens but this split-lens is adjusted to more pieces rather than the simple two pieces. The generated N-lens sectors are all separated the same distance to the center and share the same focal length. Therefore, we obtain diverse focal points drawing a circle at the focal plane. In particular, if the number of sectors is sufficient enough, a continuous split-lens configuration is created, which leads to a light ring at the focal plane. The configuration of such scheme is depicted in Figure 3.

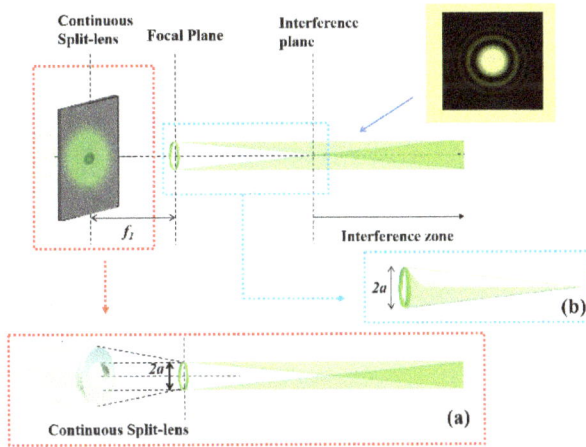

Figure 3. The light trapping scheme based on the split-lens configuration: (**a**) Continuous split-lens configuration; (**b**) Light cone structure created by the LCoS generated continuous split-lens.

In Figure 3, the input light modulated by the continuous split-lens phase distribution is propagated to the focal plane and the complete light ring (green ring) is obtained. Note that the continuous split-lens digital scheme is analogous to a classical lens continuously split and displaced from its center, as depicted in Figure 3a. In addition, a detail of the light cone trapping structure created by using this scheme is depicted in Figure 3b. In particular, light propagation after the light ring creates a hollow light cone that can be used as a basic structure for particle trapping. After the created light cone, interference between beams coming from different positions at the light ring occurs, and a Bessel beam structure is obtained (see the upper right corner in Figure 3).

To achieve the above-discussed light cone structure, the particular phase distribution to be sent at the LCoS display is given by Equation (4),

$$U(r, \theta) = \exp[i\frac{\pi}{\lambda f}(r - a(\theta))^2]$$ (4)

where λ is the wavelength of the input light and f is the focal length shared by all the split-lens. Moreover, we define the phase distribution in the polar coordinate system (r, θ). Finally, we use $a(\theta)$ to describe the distance from the split lens to the LCoS coordinate origin.

By using the above-described scheme, we can effectively adjust some light cone characteristics (i.e., the cone length or its spatial position) by simply changing the focal length of the split-lens or the separation distance between the split-lens sectors. In particular, the length of the created light cones can be calculated by using Equation (5), according to geometrical optics relations,

$$L = a(\theta)f/\phi$$ (5)

where ϕ is the diameter of the generated continuous lens without being spilt (the lens aperture).

At this moment, we explain how to trap microparticles inside the light cone. By using the light cone shown in Figure 3, the particle cannot fall into the light cone because of its upper surface also offers photophoretic forces that bounces the particle away before entering into the light cone central space. This is demonstrated in Figure 4a. Therefore, we have to open the light cone structure from the upper section to allow particles entering its central area. Note that to open the light cone means removing light from the upper section. This is practically reliable just by properly multiplying an angular sector to the continuous split-lens phase distribution to be sent at the LCoS display. The angular sector presents a constant phase value and therefore, light illuminating this section on the LCoS display is not being modulated according to the split-lens scheme. Under this scenario, an opening is obtained in the light cone (see Figure 4b). Once the particle enters the light cone through the upper opening, we close the light cone from the top once again, by simply removing the added angular sector and recovering the original split-lens phase distribution. After this process, the microparticle is located into the light cone (see Figure 4c). Nevertheless, we still face one major problem as even though the particle is kept steady in the vertical direction, it is still not stable within the horizontal direction, as it can escape from the light cone front side (i.e., from the light circle plane, see Figure 4c). Thus, this section has to be closed to prevent particles from escaping. This is achieved by multiplexing a regular lens together with the continuous split-lens scheme. By choosing a focal length, for the regular lens, shorter than the distance between the LCoS and the light ring plane, light modulated by the regular lens acts as an stopper at the cone opened side (see effect of the regular lens represented by red light in Figure 4d). At this moment, a more sophisticated light structure, demonstrated as an optical bottle architecture, is realized within the combination of these two diffractive elements (regular lens and continuous split-lens). Under this scenario, the particle inside the light cone is now stably trapped within the new optical bottle as both sides are sealed. This optical bottle structure is demonstrated in Figure 4d. Last but not least, by simultaneously adjusting the focal lengths for both light structures, the whole optical bottle structure can be spatially moved along the axial direction. By doing this, the trapped microparticle is dragged by the light system, and the spatially manipulation of microparticles is efficiently realized.

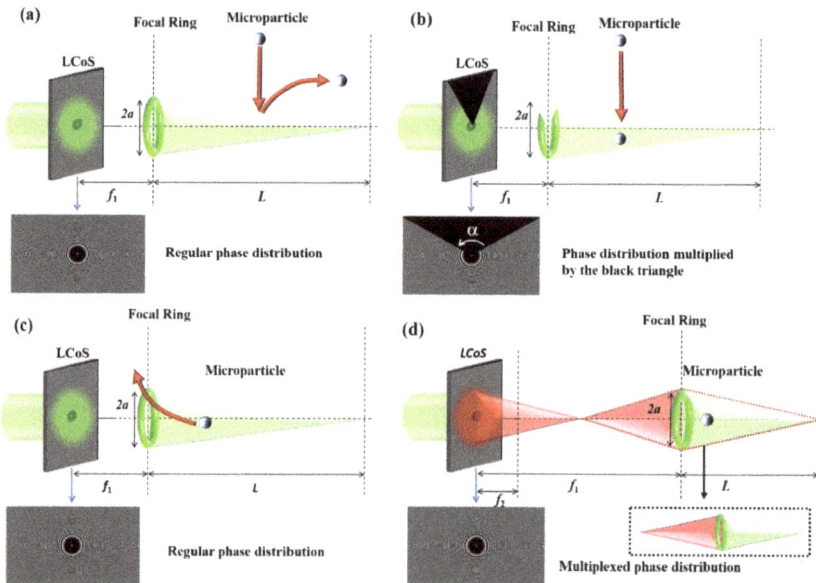

Figure 4. Sketch of the proposed microparticle trapping method: (**a**) Microparticle being bounced when encountering a closed light cone; (**b**) Microparticle entering the light cone with its upper section opened; (**c**) Microparticle escaping from the light cone thorough the unsealed section; (**d**) Microparticle trapped in the sealed light capsule structure.

2.3. LCoS Display Based Inline Holographic System

In this section we introduce an inline (IL) holographic system to monitor particles in arbitrary spatial positions [23,24]. When performing holographic imaging of objects, the conjugate image (also referred as ghost image) is always added to the final image when dealing with inline holographic architectures, this situation dramatically deteriorating the image quality. To avoid this, off-axis set-ups for holographic imaging have been proposed in literature [33,34]. However, IL set-ups present some benefits when compared with off-axis set-ups, for instance, inline holographic systems provide a larger spatial bandwidth than off-axis systems [35].

An interesting proposal was given in Ref. [24,29], where the sideband technique was proposed for holographic imaging of objects, being able to remove ghost images into an IL system. The idea was to use a filter in the Fourier space to remove the contribution of the conjugate image. However, this filtering also removed object content, and the final image was degraded. We present in this section a double-sideband filter IL holographic system from which the undesired conjugate image is eliminated. More importantly, the full content of the object is retrieved, almost at real time, so it can be used for dynamic processes. To achieve the conjugate image removing, we implement the LCoS display as the digital double shutter.

To perform the double-sideband filtering, we firstly consider an almost transparent object (i.e., the microparticle with an extremely small diameter) and therefore, the wavefront passing through this object can be written as Equation (6),

$$U_o(x,y) = 1 + \Delta U_o(x,y), \tag{6}$$

where $\Delta U_o(x,y)$ refers to the small object amplitude variations. Thus, the Fourier transform of the wavefront amplitude distribution can be calculated as Equation (7),

$$\widetilde{U}_o(\mu,\nu) = \delta(\mu,\nu) + \Delta\widetilde{U}_o(\mu,\nu), \tag{7}$$

where δ represents the Dirac delta function, μ and ν refer to the spatial frequencies. The symbol ~ refers to the Fourier space.

Later, by implementing one section of the double-sideband filter at the Fourier plane, which is feasible to block the $\mu < 0$ frequencies, we obtain the filtered wavefront amplitude at the image space (we refer it as CCD plane) as Equation (8) [29],

$$U_{CCD}^+ = \frac{1}{2} + \int\limits_0^\infty d\mu \int\limits_{-\infty}^\infty \Delta\widetilde{U}_o(\mu,\nu)e^{i2\pi(\mu x+\nu y)}d\nu, \tag{8}$$

The intensity can be acquired as Equation (9) by calculating the square modulus of Equation (8),

$$I_{CCD}^+ = |U_{CCD}^+|^2 \approx \frac{1}{4} + \underbrace{\frac{1}{2}\int\limits_0^\infty d\mu \int\limits_{-\infty}^\infty \Delta\widetilde{U}_o(\mu,\nu)e^{i2\pi(\mu x+\nu y)}d\nu}_{A^+}$$

$$+ \underbrace{\frac{1}{2}\int\limits_{-\infty}^0 d\mu \int\limits_{-\infty}^\infty \Delta\widetilde{U}_o^*(-\mu,-\nu)e^{i2\pi(\mu x+\nu y)}d\nu}_{B^+}, \tag{9}$$

where the fourth term which contained the small value of $|\Delta\widetilde{U}_o|^2$ is neglected considering the transparent assumption of the object.

The second term (A^+) in Equation (9) only carries the positive frequencies $\mu > 0$ of the object image, and on the contrary, the third term (B^+) contains the negative frequencies $\mu < 0$ of the conjugate image (note that asterisk * in the third term represents the complex conjugation). Finally, we digitally

Fourier transform the obtained intensity (i.e., I^+_{CCD} in Equation (9)), and the $\mu < 0$ frequencies are removed. Under this scenario, the complex amplitude is calculated as Equation (10),

$$|U^+_{CCD}|^2 \approx \frac{1}{4} + \underbrace{\frac{1}{2}\int\limits_0^\infty d\mu \int\limits_{-\infty}^\infty \Delta\widetilde{U}_o(\mu,\nu)e^{i2\pi(\mu x+\nu y)}d\nu}_{A^+}, \tag{10}$$

Note that by only addressing one sideband filter (which in fact is one section of the double-sideband filter), the undesired conjugate term (i.e., the B$^+$ term in Equation (9)) is removed. However, the complex amplitude in Equation (10) cannot guarantee a complete wavefront reconstruction as the $\mu < 0$ frequencies are missing. Therefore, a second sideband filter able to only filter the $\mu > 0$ frequencies is addressed. As will be further explained, both sideband filters are simultaneously addressed thanks to the use of the LCoS display. The intensity obtained at the CCD plane by using this second filter can be calculated in a way equivalent to the above-stated filtering process, but now, by removing $\mu > 0$ frequencies. It leads to the following expression Equation (11) [29],

$$|U^-_{CCD}|^2 = \frac{1}{4} + \underbrace{\frac{1}{2}\int\limits_{-\infty}^0 d\mu \int\limits_{-\infty}^\infty \Delta\widetilde{U}_o(\mu,\nu)e^{i2\pi(\mu x+\nu y)}d\nu}_{A^-}. \tag{11}$$

The finally object image, containing the whole spatial frequency distribution of the quasi-transparent object, can be finally written as Equation (12) by combining Equations (10) and (11),

$$I = |U^-_{CCD}|^2_{A^-} + |U^+_{CCD}|^2_{A^+} = \frac{1}{2} + \underbrace{\frac{1}{2}\int\limits_{-\infty}^0 d\mu \int\limits_{-\infty}^\infty \Delta\widetilde{U}_o(\mu,\nu)e^{i2\pi(\mu x+\nu y)}d\nu}_{A^-} + \underbrace{\frac{1}{2}\int\limits_0^\infty d\mu \int\limits_{-\infty}^\infty \Delta\widetilde{U}_o(\mu,\nu)e^{i2\pi(\mu x+\nu y)}d\nu}_{A^+} = \frac{1}{2}U_0(x,y). \tag{12}$$

In Equation (12), we can clearly distinguish that the conjugate image is eliminated and therefore the object image is obtained without the distortions associated to ghost images. Once the full complex amplitude is assured (i.e., Equation (12)), the holographic wavefront image can be reconstructed at any axial plane by using some diffraction method. We used the Rayleigh-Sommerfeld diffraction method, and the reconstructed wavefront is given by Equation (13) [36,37],

$$U(x,y,d) = \iint \widetilde{I}(\mu,\nu) \times \exp\{\frac{i2\pi}{\lambda}d[1 - \lambda^2(\mu^2 + \nu^2)]^{1/2}\} \times \exp[i2\pi(x\mu + y\nu)]d\mu d\nu, \tag{13}$$

where $\widetilde{I}(\mu,\nu)$ is the Fourier transfer of Equation (12) and d is the axial distance from the image input plane to the particular position where we want to reconstruct the wavefront.

So far, we have theoretically discussed the mathematical background to implement the IL holographic wavefront imaging through a double-sideband filter. Here, an optical scheme to illustrate how the real implementation of such system can be conducted by using an LCoS display (as the digital double-sideband filter) is presented. The optical sketch is shown in Figure 5, where a polarized collimating light is used as the illumination with its polarization direction rotated 45° to Y-axis of the world coordinate system. Afterwards, a convergent lens is inserted after the tested object and therefore, we can determine the Fourier plane at its focal plane F. By now, the LCoS display is precisely implemented at the Fourier plane so that it can provide the frequency filtering. Even though that the spatial light modulator presented in Figure 5 is described as a transmissive element for the sake of clarity, in fact, it is the LCoS display working within the reflective scheme. The actual reflective configuration is conducted by combining the LCoS display with a beam splitter (it is also shown in Figure 5; see red dashed inset image). Into this LCoS display, we set two sectorial phase retardations: half of the display is driven to a phase of 0° and the other half to 180° (see Figure 5). Under this scenario, the polarization direction of the input light passing through the LCoS is rotated for 90°

in the upper section (black section, δ as 180°), whereas on the contrary, the polarization direction corresponding to the other section (white section, δ as 0°) is maintained.

Later, two linear analyzers (LP1 and LP2) are set before the two CCDs cameras. Their transmissive directions are set perpendicular to each other: LP2 is oriented 45° to the Y-axis (the same polarization direction as the input light) and LP1 is oriented to 135° to the Y-axis. Under this scenario, the combination of the LCoS polarization modulation and the analyzers LP1 and LP2 orientations acts as the double sideband filter. Light recorded by CCD1 was previously filtered as half of the object frequencies are blocked. This can be assimilated to Equation (10). On the contrary, LP2 acts as a second filter and the wavefront recorded by the second CCD can be assimilated to Equation (11). Therefore, the final reconstructed holographic wavefront imaging of the object without the conjugate image can be achieved by considering Equation (13). Moreover, we want to note that such IL holographic system presents another advantage apart from the conjugate image removing as it is feasible to record dynamic processes, which is extremely favored for the study of particles in motion.

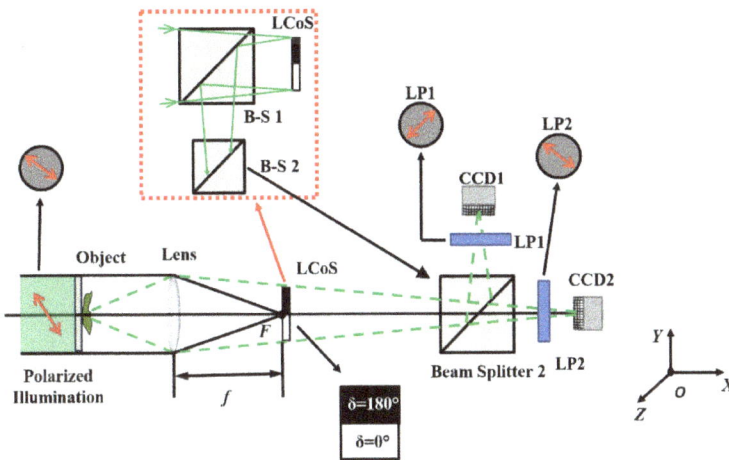

Figure 5. LCoS display based double-sideband filter IL holographic system.

3. Results

The experimental implementation of the LCoS self-calibration is preliminarily presented in this section using the split-lens configuration as well the micro-lens array scheme (Section 3.1). Afterwards, in Section 3.2 we show the experimental particle trapping and manipulation, by using the method described in Section 2.2. Finally, in Section 3.3 we show the experimental implementation of the inline holographic system proposed in Section 2.3, and the practicability is verified by monitoring some dynamic microparticles.

3.1. LCoS Display Self-Calibration

In this subsection, we present the experiments to realize both the LCoS display phase-voltage and surface profile calibration.

3.1.1. Phase-Voltage Calibration

In this section, the method described in Section 2.1.1, is experimentally implemented and tested to determine the phase-voltage relation of the LCoS display. The sketch shown in Figure 1 is experimentally implemented and shown in Figure 6. We use a polarized He-Ne laser with the wavelength of 632.8 nm as the illumination. Later, light is expanded and collimated through the combination of a microscope objective (MO), a pinhole and a convergent lens. Note that the combination of such three elements not only makes light expansion feasible, but also they guarantee light filtering. A half-waveplate (HWP) combined with a linear polarizer (LP) is used to control the intensity of light illuminating the LCoS. In addition, the direction of LP is selected to ensure an LCoS display performance in the phase-only regime (i.e., the polarization direction is parallel to the LC director direction). Once the input light is properly modulated, the PLUTO-LCoS display (distributed by HOLOEYE) is aligned with its surface strictly perpendicular to the input beam (see the zoomed figure at the right bottom corner of Figure 6). Here, the LCoS display presents a 1920 × 1080 resolution with the pixel size as 8 μm, and it has the filling factor as 87%. Finally, the collimated light entering the LCoS display is modulated by the digital two split-lens configuration (see Figure 1a) and the corresponding fringes-like interference pattern is obtained at the CCD camera, with the help of the beam splitter (B-S) and the mirror placed after the LCoS.

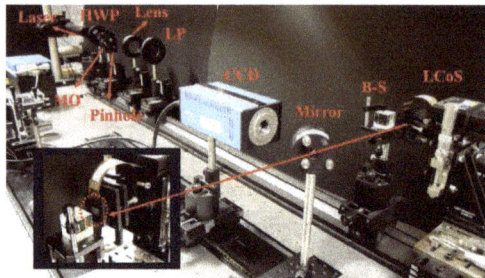

Figure 6. LCoS self-calibration implementation.

Once the system set-up is determined, the two split-lens configuration is addressed to the LCoS display. The focal length of the addressed split-lens is set to 350 mm and the separation distance as 0.4 mm. Later, the two split-lens scheme is addressed for different values of the constant phase $\phi(V)$, so that the interferometry pattern is modified (see Section 2.1). To realize the whole gray level range calibration, we drive different voltages (phases) by the means of changing the gray level from 0 to 255 with the step of 8. Therefore, 33 different interferometry patterns are obtained in the CCD with each pattern relates one phase value (voltage) to its corresponding gray level. As stated in Section 2.1, from this collection of interference patterns, the phase-voltage look-up table can be retrieved [22]. The obtained experimental curve, representing the phase as the function of the gray level, is shown in Figure 7. Each data in the curve given in Figure 7 is obtained by repeating the experiment for one hundred times, hence we also calculate the corresponding standard deviations, which are presented as the error bars in red. From Figure 7, we see that the phase-gray level curve demonstrates a nearly linear distributed tendency and the error bar values along the whole distribution are small. What is more, the phase values after the calculation is ranging from 0 to ~6.28 radians, which is suitable for phase-only applications.

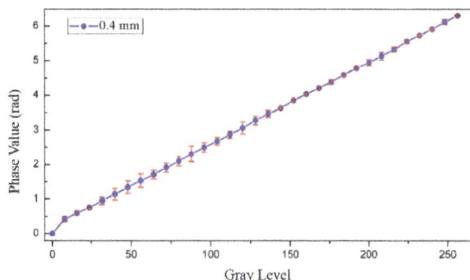

Figure 7. Phase-gray level curve measured using the split-lens separation distance as 0.4 mm.

3.1.2. Surface Homogeneity Calibration

The surface homogeneity calibration shares the same experimental set-up given in Figure 6, and it is achieved just by changing the phase pattern addressed to the LCoS display. In this case, the Shack–Hartmann configuration is addressed (see Section 2.1.2). In particular, we generate a 4×2 micro-lens pattern in which each micro-lens scheme occupies 400×400 LCoS pixels. Therefore, the whole micro-lens scheme presents an LCoS display dimension as 1600×800 pixels. The phase distribution of such scheme is given in Figure 8a and the corresponding spot-array received at the focal plane is given as Figure 8b. Note that such eight micro-lenses pattern cannot cover the whole LCoS surface, not to mention so few focal points cannot guarantee a precise surface integration [31]. Therefore, we dynamically displace the generated micro-lens pattern along the whole LCoS screen both in x and y directions. In both directions, this is implemented for 8 times with a step of 50 pixels. Finally, by recording the 8 focal points within each step (64 displacements in total) we can construct a full image with 512 intensity spots in the CCD (see Figure 8c). Once the final spot array image is obtained, we implement a 38×38 reference grid to analyze the spot deviations from their corresponding centers.

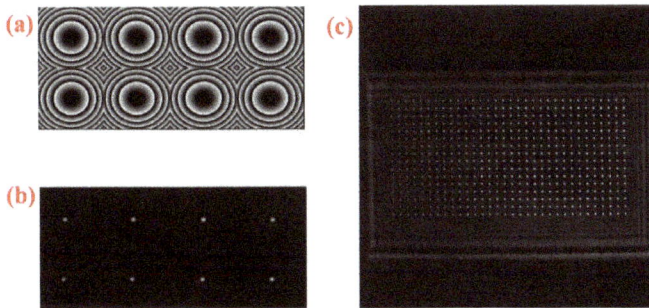

Figure 8. (a) Phase distribution sent to the LCoS display to generate the 4×2 diffractive micro-lens; (b) the intensity spot array at the focal plane obtained by illuminating the 4×2 diffractive micro-lens with collimated light; and (c) the final 512 intensity spots image obtained by performing micro-lens array displacements.

As detailed in Section 2.1.2, by calculating the deviations of light dots in the final spot array image (i.e., image in Figure 8c) from a reference pattern, the final LCoS surface can be reconstructed. Obtained results are given in radian units in Figure 9a, where a quadratic profile is clearly observed. This quadratic deformation is normally introduced by the lateral mechanical stress applied on the screen during the fabrication process, when the LC surface is stuck to the bottom structures of the LCoS display. To describe the quadratic profile of the LCoS display quantitatively, we studied the pixel-value cross section of a horizontal line in Figure 9a (from the middle pixel of the surface left edge to the middle pixel of the surface right edge). The corresponding peak-to-valley (PV) value of such line is 28.01 radians. To compensate such quadratic deviation, we implement a further step by adding the inverse phase distribution of Figure 9a to the LCoS display. Under this scenario, we repeated the LCoS surface measurement, once again by using the same self-addressed S-H scheme. Ideally, the screen profile after such compensation should be flat. In reality, the measured compensated surface profile after correction is given in Figure 9b, where we can easily find the quadratic error is almost eliminated, and the obtained surface is smoother. In particular, the surface PV (for the same pixels line previously stated) after correction is decreased to just 1.28 radians.

LCoS display surface profile before compensation   LCoS display surface profile after compensation

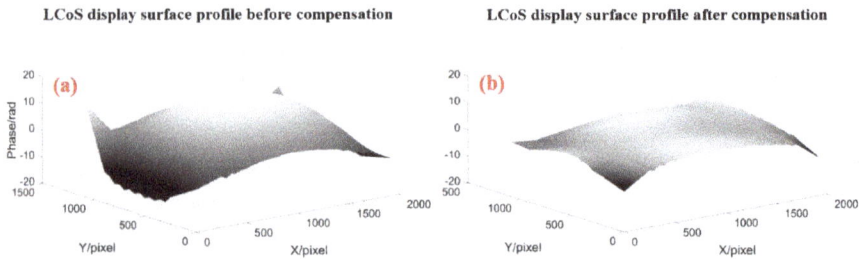

Figure 9. Surface profiles of the measured LCoS display: (**a**) LCoS display surface profile before compensation; and (**b**) LCoS display surface profile after compensation.

3.2. Microparticle Manipulation through the LCoS Display Generated Split-Lens Schemes

In this subsection, we implement the method described in Section 2.2 to create a customized three dimensional light capsule to capture and spatially manipulate microparticles. The optical set-up implemented to such aim is given in Figure 10.

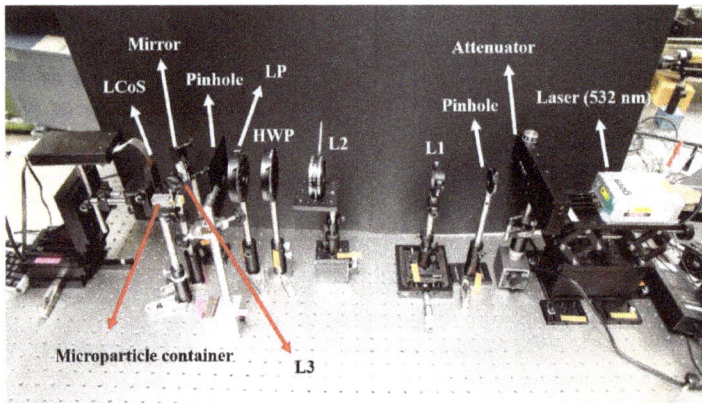

Figure 10. LCoS display based microparticle manipulation system.

The system is illuminated with a 532 nm wavelength laser (gem 532 Model, Laser Quantum) with the output power ranging from 50 mW to 2 W. Afterwards, one attenuator combined with a pinhole is inserted into the system to properly adjust the intensity. Then, light passing through an afocal system, formed by a divergent lens (L1, f_1 as -100 mm) and a convergent lens (L2, f_2 as 300 mm), is collimated and expanded. Next, the LCoS display is located perpendicular to the input beam. A linear polarizer (LP) placed before the LCoS is properly oriented to ensure a phase-only performance of the device. Under this scenario, we send two multiplexed phase distributions to the LCoS display: (i) the first one defines the continuous split-lens configuration (creates the light cone); and (ii) the second stands for a regular lens (configures the cone cork). See more details in explanation related to Figure 4. Specifically, the parameters set to implement the continuous split-lens are: the distance between sector centers to the origin of $a = 1.07$ mm, a focal length of $f = 370$ mm and an aperture of $\phi = 2.07$ mm (see Equations (4) and (5)). Conversely, the focal length for the regular lens is set 350 mm. Note that in this scheme, the focal length of the regular lens (i.e., 350 mm) is smaller than the continuous split-lens (i.e., 370 mm) which satisfies the structure described in Figure 4d. Finally, a convergent lens (L3) images the whole structure into a transparent container, in which a number of microparticles are stuck to the top face due to electrostatic forces. By taking into account all these experimental parameters, we have obtained an experimental optical bottle with an axial length of ~6.5 mm. Now, this compacted optical bottle is feasible to capture particles through photophoretic forces. The used transparent container is a square UV fused quartz cuvettes (Thorlab CV10Q3500) with its size of 45–10–10 mm and two sides polished. On the other hand, the used particles are carbon coated hollow glass microspheres (Cospheric HGMS-0.14 63–75 μm) with their diameters ranging from 63–75 μm. Moreover, the particle density is 0.14 g/cm^3 and the mass ranged from 18.3–30.9 ng. Here, we want to note that apart from the carbon coated microsphere, the pure carbon particles [36] or the absorbing silicon particles [37] are also feasible to be trapped. As told before, particles are preliminarily stuck to the container inner surface by electrostatic forces. At this moment, these particles are not able to enter into the optical bottle structure because light capsule is sealed. The central cross section figure corresponding to this light capsule structure is given in Figure 11a. However, optical bottle can be opened from the top by multiplying an azimuthal sector with a constant phase within the continuous lens phase distribution, as detailed in Section 2.2. As a consequence, the resulting central cross section is provided in Figure 11b, this scheme enabling particles to enter into the light structure. By softly tapping the above side of the container, we forced particles to land into the capsule and they can be trapped by photophoretic forces exerted by the light structure. Finally, we closed the upper section of the light structure once again, by removing the triangle gap, and the complete optical bottle was recovered (see Figure 11c). Once the optical bottle is sealed, the particle is stably contained into this light structure. The particle trapped in this optical bottle is demonstrated in Figure 12a where we highlight the spatial particle position with a red circle.

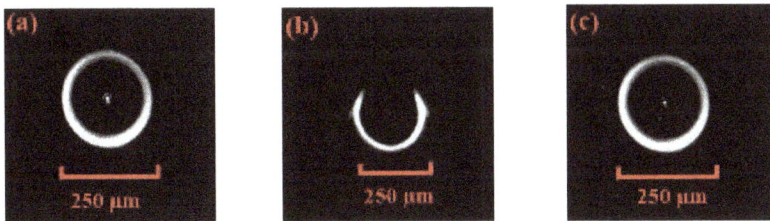

Figure 11. Central cross sections of the optical bottle structure: (**a**) the capsule structure without opening the upper section; (**b**) the capsule structure with the upper section opened by removing a triangular section of the phase distribution; (**c**) the optical bottle structure with its upper section being sealed.

Apart from the fact that we are able to trap particles in the generated optical bottle, we also moved a step further by implementing the particle manipulation. This is achieved by changing the spatial position of the optical bottle itself. To be more specific, we simply changed the focal length of the digital split-lens from 370 mm to 381 mm, and we also modified the focal length of the digital regular lens from 350 mm to 361 mm. By taking into account Equation (5), the optical bottle length was correspondingly changed from the original ~6.5 mm to the current ~5.35 mm. Moreover, due to the modification of the focal length values, the optical bottle was spatially shifted for ~0.47 mm in the optical axis direction. As the optical bottle is moved, the particle trapped inside is spatially dragged in the same shifting direction. Therefore, we can control the axial position of the particle. The axially shifted particle is demonstrated in Figure 12b, where the displacement can be discovered from the spatial difference compared to Figure 12a (i.e., the original trapped particle).

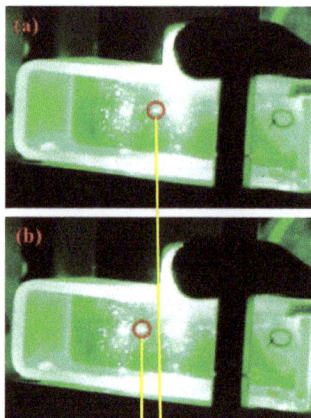

Figure 12. Microparticle trapping and spatial manipulation: (**a**) the original trapped particle; (**b**) new axial position of the trapped particle when displacing the light capsule structure.

3.3. LCoS Display Based Inline Holographic System

In this section, we present the experimental implementation to achieve the IL holographic imaging using the LCoS display (see Section 2.3 and Figure 5). The experimental implementation is shown in Figure 13. We use as illumination a 633 nm wavelength linear polarized laser with the output power as 17 mW. The laser is then spatially filtered and later collimated by a convergent lens L1 (focal length as 250 mm). The object to be studied is placed to a distance after L1. Then, we insert a second convergent lens with a focal length of 300 mm (L2, see Figure 13). The function of L2 is twofold: one the one hand, it images a plane of the object space (we label it as object plane *P*) into the CCD1 and CCD2 camera planes (the object plane was situated at 690 mm before L2); on the other hand, the Fourier spectrum of the object is set at the focal plane of L2, where the LCoS display is set. Note that this configuration accomplishes the scheme shown in Figure 5.

As the LCoS used (distributed by HOLOEYE) works into a reflective configuration, a system of beam-splitters (B-S1 and B-S2 in Figure 13) are used to properly imaging the object to CCD1 and CCD2. What is more, the obtained images at the CCDs are filtered according to the double-sideband filter explained in Section 2.3. This is experimentally achieved thanks to combination of the LCoS display with two linear polarizers (LP1 and LP2 in Figure 13), which are placed in front of the CCD1 and CCD2 cameras, respectively. As CCD cameras we used two Basler CCDs (KAI-1020) which guarantee the resolution of 1MP with the frame rate as 60 fps. Note that for a better vision, the optical elements before the second convergent lens (i.e., the laser, the primary convergent lens and the spatial filter) are ignored in Figure 13.

By computational processing the images recorded by CCD1 and CCD2, according to the mathematical formulation described in Section 2.3, holographic images of objects can be retrieved without the degradation associated to ghost images. In particular, we implemented two objects for the measurement at the same time: a reticle and a thin glass plate. In the case of the thin glass plate, it is covered with microparticles with sizes of ~100 μm. The reticle is located 50 mm before the *P* plane (i.e., 740 mm before L2) and the thin glass plate with microparticles is implemented closer to L2, 50 mm behind the *P* plane (i.e., 640 mm before L2). In this way we can prove how we simultaneously image the reticle and the microparticles, which are at different axial planes. Moreover, the thin glass plate is hinged to a rotation mechanism from which we can provide the dynamic potential of the method.

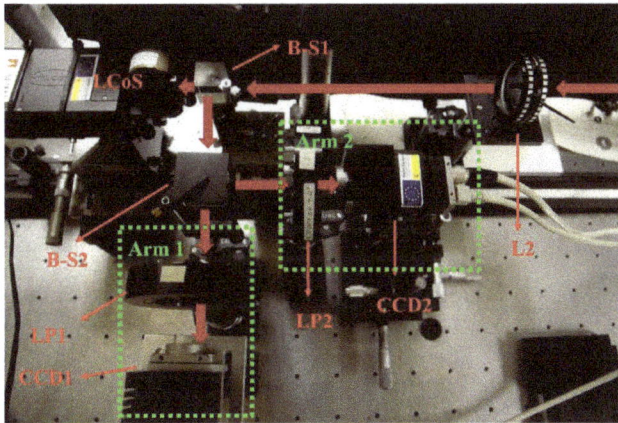

Figure 13. Optical set-up of the IL holographic system.

By applying the method explained in Section 2.3 we can reconstruct the complex amplitude at the object plane *P*. From this information, reticle and thin glass plate can be also retrieved by properly applying the Rayleigh-Sommerfeld diffraction method (Equation (6)) [38,39]. In particular, by taking into account the magnification introduced by L2, the reticle plane is reconstructed by setting the reconstruction position of $z = -30$ mm to the object plane *P* (note *P* is set at $z = 0$). Results are given in Figure 14a, where the image of the reticle is clearly distinguished but the microparticles are defocused (see Figure 14a). Afterwards, by changing the reconstruction position to $z = +31$ mm, we focus the microparticles plane (microparticles on the thin glass plate) but the reticle at this time is defocused (see Figure 14b). Such pattern shifting demonstrates the feasibility of using the IL holography system to realize the object imaging at different axial positions. Finally, we rotated as well the thin glass plate from which we observed the microparticle dynamic rotation. The rotated microparticles are presented in Figure 14c from which the position shifting compared to Figure 14b is clearly demonstrated.

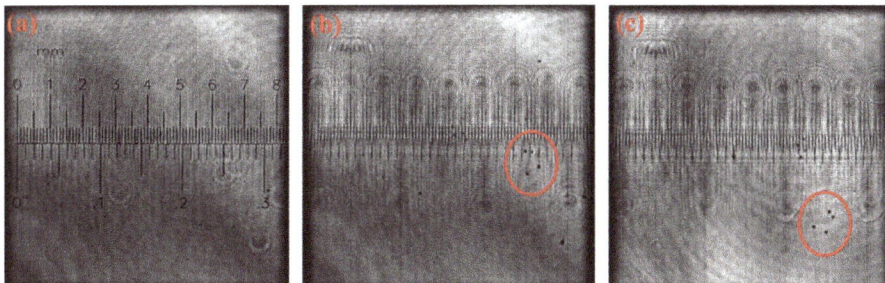

Figure 14. Images of the reticle and the microparticles captured by the IL holography system: (**a**) the holographic image of the reticle; (**b**) the holographic image of the microparticles located on the thin glass plate (see the red circle); and (**c**) the holographic image of the rotated microparticles located on the thin glass plate (see the red circle).

4. Conclusions

We reviewed different techniques we recently proposed to calibrate and apply LCoS display technology. We preliminarily use a self-calibration method to determine the phase-gray level relation and the surface profile of an LCoS display. In particular, the phase-voltage relation is experimentally determined by using the diffractive split-lens configuration. Such method leads to an interferometric pattern that is transversally displaced as a function of the gray level. From this relation, the phase-gray level curve is accurately determined. On the other hand, by using the same optical arrangement, but addressing a different self-addressed hologram (in this case, the Shack–Hartmann scheme), we are able to determine the LCoS screen profile. From this information, the effect of the screen distortions was corrected, reaching a higher performance of the device. Afterwards, this calibrated LCoS display was implemented into an optical set-up able to trap and manipulate microparticles. To this aim, three-dimensional light structures were created by using split-lens based configurations. The created structures configured an optical bottle, where some particles (glass coated microspheres) were experimentally trapped. Moreover, by simply modifying some control parameters, we conducted the spatial shifting of the optical bottle in the axial direction, which allowed us to control the spatial position of the captured particle. Finally, a method to obtain holographic images of objects was also described. This is an inline (IL) holographic system based on an LCoS display. The method shares all the benefits of inline schemes, but avoids one major problem associated to IL systems, the non-desired influence of ghost images. This is achieved by implementing a double-sideband filter (DSF) at the Fourier plane of the object. The DSF is implemented thanks to the combination of the LCoS display with a pair of linear analyzers. We experimentally demonstrated how the method is able to eliminate the unsatisfied conjugate image. What is more, this LCoS based IL holographic scheme avoids the necessity of time-sequential measurements and therefore, we achieved the dynamic holographic observation.

Author Contributions: Conceptualization, H.Z., A.L. and J.C.; writing—original draft preparation, H.Z.; writing—review and editing, A.L.; software, J.C.; experimental work, H.Z., A.V.E., and C.R.; methodology, A.T., C.I.; data analysis, H.Z.; supervision, J.C. and A.L.

Funding: Spanish MINECO (FIS2015-66328-C3-1-R and fondos FEDER); Catalan Government (SGR 2014-1639). Chinese Scholarship Council (201504910783).

Conflicts of Interest: The authors declare no conflict of interest.

References

1. Mu, Q.; Cao, Z.; Hu, L.; Li, D.; Xuan, L. Adaptive optics imaging system based on a high-resolution liquid crystal on silicon device. *Opt. Express* **2006**, *14*, 8013–8018. [CrossRef] [PubMed]
2. Fernández, E.J.; Prieto, P.M.; Artal, P. Wave-aberration control with a liquid crystal on silicon (LCOS) spatial phase modulator. *Opt. Express* **2009**, *17*, 11013–11025. [CrossRef] [PubMed]
3. Zhang, Z.; You, Z.; Chu, D. Fundamentals of phase-only liquid crystal on silicon (LCOS) devices. *Light Sci. Appl.* **2014**, *3*, e213. [CrossRef]
4. Matsumoto, N.; Itoh, H.; Inoue, T.; Otsu, T.; Toyoda, H. Stable and flexible multiple spot pattern generation using LCOS spatial light modulator. *Opt. Express* **2014**, *22*, 24722–24733. [CrossRef] [PubMed]
5. Marquez, A.; Moreno, I.; Iemmi, C.; Lizana, A.; Campos, J.; Yzuel, M.J. Mueller-Stokes characterization and optimization of a liquid crystal on silicon display showing depolarization. *Opt. Express* **2008**, *16*, 1669–1685. [CrossRef] [PubMed]
6. Ni, J.; Wang, C.; Zhang, C.; Hu, Y.; Yang, L.; Lao, Z.; Xu, B.; Li, J.; Wu, D.; Chu, J. Three-dimensional chiral microstructures fabricated by structured optical vortices in isotropic material. *Light Sci. Appl.* **2017**, *6*, e17011. [CrossRef] [PubMed]
7. Kowalczyk, A.P.; Makowski, M.; Ducin, I.; Sypek, M.; Kolodziejczyk, A. Collective matrix of spatial light modulators for increased resolution in holographic image projection. *Opt. Express* **2018**, *26*, 17158–17169. [CrossRef] [PubMed]
8. Fuentes, J.L.M.; Moreno, I. Random technique to encode complex valued holograms with on axis reconstruction onto phase-only displays. *Opt. Express* **2018**, *26*, 5875–5893. [CrossRef] [PubMed]
9. Vinas, M.; Dorronsoro, C.; Radhakrishnan, A.; Benedi-Garcia, C.; Lavilla, E.A.; Schwiegerling, J.; Marcos, S. Comparison of vision through surface modulated and spatial light modulated multifocal optics. *Biomed. Opt. Express* **2017**, *8*, 2055–2068. [CrossRef] [PubMed]
10. Qaderi, K.; Leach, C.; Smalley, D.E. Paired leaky mode spatial light modulators with a 28° total deflection angle. *Opt. Lett.* **2017**, *42*, 1345–1348. [CrossRef] [PubMed]

11. Cheng, Q.; Rumley, S.; Bahadori, M.; Bergman, K. Photonic switching in high performance datacenters [Invited]. *Opt. Express* **2018**, *26*, 16022–16043. [CrossRef] [PubMed]

12. Lizana, A.; Zhang, H.; Turpin, A.; Van Eeckhout, A.; Torres-Ruiz, F.A.; Vargas, A.; Ramirez, C.; Pi, F.; Campos, J. Generation of reconfigurable optical traps for microparticles spatial manipulation through dynamic split lens inspired light structures. *Sci. Rep.* **2018**, *8*, 11263. [CrossRef] [PubMed]

13. Zhang, H.; Lizana, A.; Van Eeckhout, A.; Turpin, A.; Iemmi, C.; Márquez, A.; Moreno, I.; Torres-Ruiz, F.A.; Vargas, A.; Pi, F.; et al. Dynamic microparticle manipulation through light structures generated by a self-calibrated Liquid Crystal on Silicon display. *Proc. SPIE* **2018**, *10677*, 106772O. [CrossRef]

14. Fuentes, J.L.M.; Fernández, E.J.; Prieto, P.M.; Artal, P. Interferometric method for phase calibration in liquid crystal spatial light modulators using a self-generated diffraction-grating. *Opt. Express* **2016**, *24*, 14159–14171. [CrossRef] [PubMed]

15. Zhang, H.; Lizana, Á.; Iemmi, C.; Monroy-Ramirez, F.A.; Márquez, A.; Moreno, I.; Campos, J. LCoS display phase self-calibration method based on diffractive lens schemes. *Opt. Lasers Eng.* **2018**, *106*, 147–154. [CrossRef]

16. Zhang, H.; Lizana, A.; Iemmi, C.; Monroy-Ramírez, F.A.; Marquez, A.; Moreno, I.; Campos, J. Self-addressed diffractive lens schemes for the characterization of LCoS displays. *Proc. SPIE* **2018**, *10555*, 105550I. [CrossRef]

17. Lizana, A.; Vargas, A.; Turpin, A.; Ramirez, C.; Estevez, I.; Campos, J. Shaping light with split lens configurations. *J. Opt.* **2016**, *18*, 105605. [CrossRef]

18. Cofré, A.; Vargas, A.; Torres-Ruiz, F.A.; Campos, J.; Lizana, A.; Sánchez-López, M.M.; Moreno, I. Dual polarization split lenses. *Opt. Express* **2017**, *25*, 23773–23783. [CrossRef] [PubMed]

19. Lobato, L.; Márquez, A.; Lizana, A.; Moreno, I.; Iemmi, C.; Campos, J. Characterization of a parallel aligned liquid crystal on silicon and its application on a Shack-Hartmann sensor. *Proc. SPIE* **2010**, *7797*, 77970Q. [CrossRef]

20. Schwiegerling, J.; DeHoog, E. Problems testing diffractive intraocular lenses with Shack-Hartmann sensors. *Appl. Opt.* **2010**, *49*, D62–D68. [CrossRef] [PubMed]

21. López-Quesada, C.; Andilla, J.; Martín-Badosa, E. Correction of aberration in holographic optical tweezers using a Shack-Hartmann sensor. *Appl. Opt.* **2009**, *48*, 1084–1090. [CrossRef] [PubMed]

22. Lizana, A.; Moreno, I.; Marquez, A.; Iemmi, C.; Fernandez, E.; Campos, J.; Yzuel, M.J. Time fluctuations of the phase modulation in a liquid crystal on silicon display: Characterization and effects in diffractive optics. *Opt. Express* **2008**, *16*, 16711–16722. [CrossRef]

23. Ramírez, C.; Lizana, A.; Iemmi, C.; Campos, J. Method based on the double sideband technique for the dynamic tracking of micrometric particles. *J. Opt.* **2016**, *18*, 065603. [CrossRef]

24. Ramirez, C.; Lizana, A.; Iemmi, C.; Campos, J. Inline digital holographic movie based on a double-sideband filter. *Opt. Lett.* **2015**, *40*, 4142–4145. [CrossRef] [PubMed]

25. Jovanovic, O. Photophoresis—Light induced motion of particles suspended in gas. *J. Quant. Spectrosc. Radiat. Transf.* **2009**, *110*, 889–901. [CrossRef]

26. Barry, J.F.; McCarron, D.J.; Norrgard, E.; Steinecker, M.H.; DeMille, D. Magneto-optical trapping of a diatomic molecule. *Nature* **2014**, *512*, 286–289. [CrossRef] [PubMed]

27. Maunz, P.; Puppe, T.; Schuster, I.; Syassen, N.; Pinkse, P.W.; Rempe, G. Cavity cooling of a single atom. *Nature* **2004**, *428*, 50–52. [CrossRef] [PubMed]

28. Ashkin, A.; Dziedzic, J.M.; Yamane, T. Optical trapping and manipulation of single cells using infrared laser beams. *Nature* **1987**, *330*, 769–771. [CrossRef] [PubMed]

29. Zhang, H.; Monroy-Ramírez, A.F.; Lizana, A.; Iemmi, C.; Bennis, N.; Morawiak, P.; Piecek, W.; Campos, J. Wavefront imaging by using an inline holographic microscopy system based on a double-sideband filter. *Opt. Lasers Eng.* **2019**, *113*, 71–76. [CrossRef]

30. Cheng, C.; Chern, J. Symmetry property of a generalized billet's n-split lens. *Opt. Commun.* **2010**, *283*, 3564–3568. [CrossRef]

31. Yaroslavsky, L.P.; Moreno, A.; Campos, J. Frequency responses and resolving power of numerical integration of sampled data. *Opt. Express* **2005**, *13*, 2892–2905. [CrossRef] [PubMed]

32. Cheng, C.; Chern, J. Quasi bessel beam by billet's n-split lens. *Opt. Commun.* **2010**, *283*, 4892–4898. [CrossRef]

33. Cuche, E.; Marquet, P.; Depeursinge, C. Spatial filtering for zero-order and twin-image elimination in digital off-axis holography. *Appl. Opt.* **2000**, *39*, 4070–4075. [CrossRef] [PubMed]

34. Hong, J.; Kim, M.K. Single-shot self-interference incoherent digital holography using off-axis configuration. *Opt. Lett.* **2013**, *38*, 5196–5199. [CrossRef] [PubMed]

35. Meng, H.; Hussain, F. In-line recording and off-axis viewing technique for holographic particle velocimetry. *Appl. Opt.* **1995**, *34*, 1827–1840. [CrossRef] [PubMed]

36. Pan, Y.-L.; Hill, S.C.; Coleman, M. Photophoretic trapping of absorbing particles in air and measurement of their single-particle Raman spectra. *Opt. Express* **2012**, *20*, 5325–5334. [CrossRef] [PubMed]

37. Zhang, Z.; Cannan, D.; Liu, J.; Zhang, P.; Christodoulides, D.N.; Chen, Z. Observation of trapping and transporting air-borne absorbing particles with a single optical beam. *Opt. Express* **2012**, *20*, 16212–16217. [CrossRef]

38. Pedrini, G.; Osten, W.; Zhang, Y. Wave-front reconstruction from a sequence of inter- ferograms recorded at different planes. *Opt. Lett.* **2005**, *30*, 833–835. [CrossRef] [PubMed]

39. Grilli, S.; Ferraro, P.; De Nicola, S.; Finizio, A.; Pierattini, G.; Meucci, R. Whole optical wavefields reconstruction by Digital Holography. *Opt. Express* **2001**, *9*, 294–302. [CrossRef] [PubMed]

![applied sciences logo] *applied sciences*

MDPI

Article

Liquid Crystal Spatial Light Modulator with Optimized Phase Modulation Ranges to Display Multiorder Diffractive Elements

Elisabet Pérez-Cabré * and María Sagrario Millán

Departament d'Òptica i Optometria, Universitat Politècnica de Catalunya, Violinista Vellsolà 37, 08222 Terrassa, Spain
* Correspondence: elisabet.perez@upc.edu; Tel.: +34-93-739-8782

Received: 13 May 2019; Accepted: 19 June 2019; Published: 26 June 2019

Abstract: A liquid crystal on silicon spatial light modulator (LCoS SLM) with large phase modulation has been thoroughly characterized to operate optimally with several linear phase modulation ranges (π, 2π, 3π, 4π, 6π, and 8π) for an intermediate wavelength of the visible spectrum ($\lambda_G = 530$ nm). For each range, the device response was also measured for two additional wavelengths at the blue and red extremes of the visible spectrum ($\lambda_B = 476$ nm and $\lambda_R = 647$ nm). Multiorder diffractive optical elements, displayed on the LCoS SLM with the appropriate phase modulation range, allowed us to deal with some widely known encoding issues of conventional first-order diffractive lenses such as undersampling and longitudinal chromatic aberration. We designed an achromatic multiorder lens and implemented it experimentally on the SLM. As a result, the residual chromatic aberration reduces to one-third that of the chromatic aberration of a conventional first-order diffractive lens.

Keywords: liquid crystal spatial light modulator; liquid crystal on silicon device; phase characterization; phase modulation; diffractive optical element; multiorder diffractive lens; harmonic lens; chromatic aberration; aberration compensation; achromatic lens

---

1. Introduction

Diffractive optical elements (DOEs) have general advantages in comparison to their refractive counterparts such as being basically flat, thin, lightweight, and inexpensive when mass-produced. However, they exhibit large chromatic aberration (e.g., in diffractive lenses, approximately one diopter of axial chromatic aberration for every three diopters of power) and frequently low diffraction efficiency [1]. Sweeney and Sommargren [1] and Faklis and Morris [2] simultaneously and independently introduced DOEs with multiwavelength optical path-length transitions between adjacent facets, called harmonic and multiorder diffractive lenses, respectively. These lenses have hybrid properties of both refractive and diffractive lenses, and they have a common focus for a number of discrete wavelengths. Sweeney and Sommargren [1] measured the modulation transfer function (MTF) when a harmonic diffractive lens was used for imaging under either monochromatic or white light illumination. They showed that a lens with 10 wavelength phase steps approached diffraction limit behavior across the visible spectrum. Faklis and Morris [2] used multiorder diffractive lenses to design an achromatic diffractive singlet. The performance of such a lens was illustrated through MTF and Strehl ratio measurements as a function of wavelength. By 1995, advances in fabrication techniques of diffractive elements allowed development of the ideas proposed in these works. Multiorder DOEs have some constraints though. For example, higher-order structures restrain their off-axis performance. Because of the greater height at the edges of each facet, an effect appears, referred to as shadowing [3,4]. Even for diffraction angles of a few degrees, the wavefront emerging from the blazed structure has small gaps introduced by the facets that affect the diffraction efficiency of the DOE [4]. This inconvenience can be overcome with the modern technology of liquid crystal (LC) displays.

The evolution of diffractive components experienced a significant advancement with the development of electrically addressed liquid crystal spatial light modulators (LC-SLM), capable of implementing a variety of programmable DOEs. Phase-only LC-SLM with a phase modulation depth larger than 2π radians is nowadays commercially available based on LC on silicon (LCoS) technology working on a reflection regime. A parallel-aligned LCoS display reaching a dynamic phase range of 4π radians for $\lambda = 454$ nm has been used to implement second-order diffraction DOEs, such as diffractive gratings and diffractive lenses, with some advantages in terms of resolution and diffraction efficiency compared to conventional first-order DOEs [5]. An LCoS-SLM initially designed to operate in infrared provides unusually large phase modulation depths in the visible band range, for instance, from 6π radians in the red region to 10π radians in the blue region [6]. This performance allows the display of a blazed diffractive grating with a reduced chromatic dispersion.

The discrete pixel structure [7] and the quantization of the available phase levels [8] of LC-SLM introduce some limitations in the implementation of DOEs on such displays. The spatial resolution of the screen may limit efficient encoding of diffractive lenses. On the one hand, low-resolution lenses implemented in the pixelated structure of LC-SLM introduce an inherent apodizing effect on the system point spread function (PSF). Apodization diminishes the secondary maxima of the PSF, but it increases the width of the central lobe [9,10]. These effects become more severe as the width of the pixel increases or as the focal length of the lens decreases. However, novel LCoS displays with a smaller pixel size and larger fill factor reduce the apodizing effect of the pixelated structure as they better approach the ideal continuous lens function. On the other hand, for diffractive lenses of low f-number, the facets become denser at the boundary of the lens aperture so the resolution needed to encode them may exceed the Nyquist frequency of the pixels on the LC-SLM [7]. To avoid aliasing of the lens function at the periphery, the design focal length f of the diffractive lens must meet the condition $f_N < f < 50 f_N$, where the inferior limit f_N is the Nyquist focal length defined in [7,11].

In this paper, we present experimental results for multiorder diffractive lenses implemented in a parallel-aligned LCoS from Holoeye that reaches up to an 8π phase modulation range in the green region of the visible spectrum. To this end, we thoroughly characterize the device to optimize the setting parameter selection so as to obtain various linear operating phase modulation ranges (π, 2π, 3π, 4π, 6π, and 8π) for an intermediate wavelength of the visible spectrum ($\lambda_G = 530$ nm). Multiorder diffractive lenses with multiwavelength jumps at the edge of their facets permit to implement, on the LCoS device, higher optical power lenses with their focal length below the Nyquist focal length. These multiorder diffractive lenses overcome the aliasing effects that encoding of common first-order diffractive lenses have. An extended characterization of the LCoS display is carried out by determining the experimental phase modulation not only for the design wavelength, $\lambda_G = 530$ nm, but also for two additional wavelengths at the extremes of the visible spectrum ($\lambda_B = 476$ nm and $\lambda_R = 647$ nm). Based on the measured chromatic performance of the LCoS screen, we design an achromatic diffractive lens that operates with higher diffraction orders. We take advantage of the larger phase depth modulation of recently available parallel-aligned LCoS modulators to implement these DOEs in such a device with the additional benefits of real-time reprogramming and no shadowing effects. Moreover, further studies of the optical properties of multiorder diffractive lenses can be conducted without the need for physically manufacturing them.

We provide the experimental results for two multiorder lenses that overcome the issue of displaying a diffractive lens of optical power beyond the Nyquist interval and an achromatic diffractive lens. We compare their performances with that of a conventional, first-order diffractive lens.

2. Multiorder Diffractive Lenses: Theoretical Background

Multiorder diffractive lenses [1–4] are commonly built so that the optical path difference (OPD) at the boundaries of adjacent zones is a multiple of λ_0, (or equivalently, a phase jump multiple of 2π rad).

Thus, $OPD = (f_0 + jq\lambda_0) - [f_0 + (j-1)q\lambda_0] = q\lambda_0$ (Figure 1), where λ_0 is the design wavelength, f_0 is the focal length when the illumination wavelength is $\lambda = \lambda_0$, and q is an integer that multiplies 2π at the phase jump. A conventional diffractive lens with facets reaching 2π phase variation (also called a modulo 2π lens) has $q = 1$.

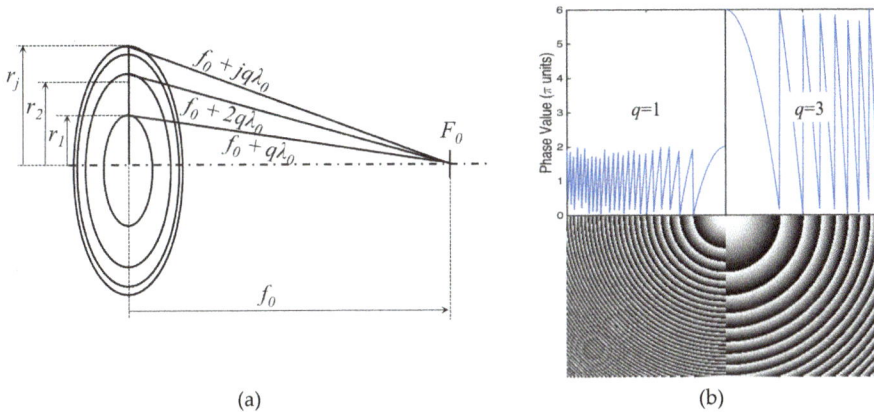

(a)   (b)

Figure 1. (a) Zone construction of a multiorder (qth-order) diffractive lens whose optical path difference (OPD) between consecutive zones is q integer times the design wavelength λ_0. (b) Profiles for a diffractive lens with 2π rad ($q = 1$) and 6π rad ($q = 3$) phase modulation depths.

The condition for obtaining a constructive interference in the focal plane F_0 of the lens determines the radial position of the jth diffractive zone, which is computed through the Pythagorean theorem

$$r_j^2 = (f_0 + jq\lambda_0)^2 - f_0^2 \approx 2jq\lambda_0 f_0. \tag{1}$$

The maximum variation of the phase profile in each zone is given by $q2\pi$. The lens has an infinite number of focal lengths according to

$$f(\lambda) = \frac{q\lambda_0 f_0}{m\lambda}. \tag{2}$$

The interested reader is referred to [2] for a detailed explanation of how Equation (2) is obtained. Note that Equation (2) justifies the highly dispersive nature of diffractive lenses.

For a conventional modulo 2π diffractive lens ($q = 1$) illuminated with the design wavelength ($\lambda = \lambda_0$), an infinite number of focal planes $f = f_0/m$ exist for the different diffraction orders [8]. If one considers a multiorder diffractive lens ($q > 1$), the optical path length $q\lambda_0$ for the design wavelength may coincide with other wavelengths for which the phase change between adjacent steps of the lens is a multiple of 2π. These wavelengths, which are called the resonant wavelengths [1], fulfill $\lambda = q\lambda_0/m$, and have a common focal plane $f(\lambda) = f_0$.

Not only does the location of the focal plane coincide for the resonant wavelengths but also the diffraction efficiency of the respective diffraction orders, which is given by [4]

$$\eta_{m,q} = \text{sinc}^2(\alpha q - m). \tag{3}$$

Coefficient α is a wavelength detuning factor that can be approximated by $\alpha = \lambda_0/\lambda$ when low-dispersive materials, or even no dispersion, are considered. The maximum diffraction efficiency $\eta_{m,q} = 1$ is achieved for resonant wavelengths at their common focal plane of a multiorder diffractive lens. For first-order diffractive lenses with $q = 1$, all the energy concentrates on the first diffraction order ($m = 1$) with a quick but smooth decrease of the efficiency for wavelengths that differ from the design wavelength. As the phase profile increases its modulation range, that is, when $q > 1$, resonant wavelengths get closer in the spectrum, and diffractive lenses of higher orders can be obtained; however, their diffracted intensity drops even faster with wavelength [4].

3. Liquid Crystal on Silicon (LCoS) Calibration Procedure and Phase Modulation Responses

We have used a parallel-aligned nematic LC device of LCoS-SLM technology from Holoeye (Pluto-BB-HR, HOLOEYE Photonics AG, Berlin, Germany) with 1920×1080 pixels of size 8×8 μm^2, fill factor of 87%, and refreshing rate of 60 Hz. This device is a phase-only SLM working on a reflective mode. An achromatic half-wave plate with a nearly flat retardance over the operation range (400–800 nm) from Thorlabs (AHWP05M-600, Thorlabs GmbH, Dachau/Munich, Germany) was used to adjust the polarization plane of the incident beam from a tunable Ar ion linearly polarized laser (35KAP431, Cvi Melles Griot, Inc., Albuquerque, New Mexico, USA) to obtain a phase-only modulation output. Normal incidence of the impinging collimated beam combined with the use of a beam splitter was chosen to assure an optimal response of the LCoS display in terms of phase modulation depth and negligible depolarization [12].

Characterization of the LCoS device was carried out to assure an optimal implementation of multiorder diffractive lenses. The procedure was done by means of the Michelson interferometer with coherent illumination shown in Figure 2. A broadband hybrid metal dielectric coating cube beamsplitter (03BSC009, Cvi Melles Griot, Inc., Albuquerque, New Mexico, USA) (BS in Figure 2) that exhibits little polarization sensitivity for the 450–700 nm spectral range was used in the setup to assure that it did not significantly affect the linear polarization state of incident beams. A flat mirror slightly tilted with respect to the SLM plane permitted us to obtain an interferogram pattern of the two beams consisting of a set of vertical fringes. The corresponding interferogram was acquired with a charged coupled device (CCD) camera (pco.1600, PCO AG, Kelheim, Germany) with 1600×1200 pixel resolution and 14 bit dynamic range (Figure 3a).

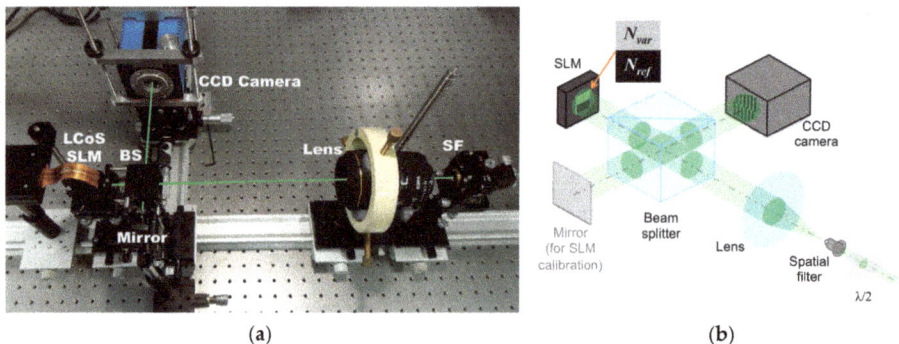

(a) (b)

Figure 2. (**a**) Michelson interferometer setup for characterization of the LCoS display (BS stands for beam splitter and SF for spatial filter). The tunable linearly polarized Ar ion-laser and the half-wave plate are not shown. (**b**) Scheme of the Michelson interferometer with a detailed view of the spatial light modulator (SLM) input plane.

The SLM screen was divided in two uniform sectors, one half with a constant grey level kept as a reference and the other half with a varying uniform grey level among the 256 grey levels available (Figure 2b). The fringe shift between the two sectors allowed us to calculate the phase variation for each grey level (Figure 3a). The procedure is described with more detail in ref. [13].

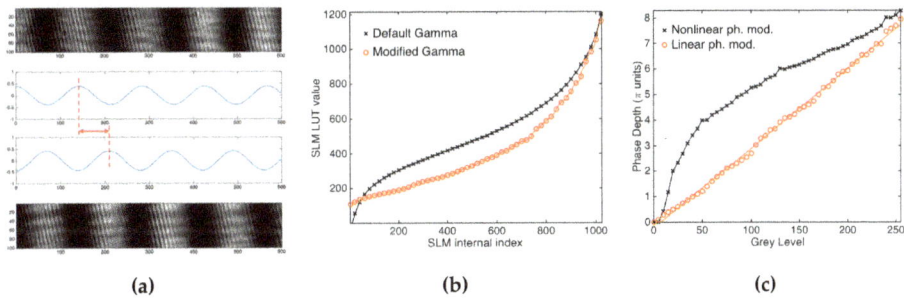

| (a) | (b) | (c) |

Figure 3. (**a**) Interferogram obtained when two different gray levels are sent to the top and bottom halves of the SLM display (Figure 2b). The phase difference is measured from the detected fringe shift. (**b**) Look-up tables (LUTs), also named gamma curves, to control the response of the SLM display and (**c**) measured phase modulation depth. Line curves with cross markers from (**b,c**) correspond to the initial measurement when the gamma curve provided by the manufacturer is used (default gamma). The corresponding phase modulation slightly exceeds 8π radians but with a nonlinear response. Line curves with circles from (**b,c**) correspond to the modified gamma curve that achieves a linear phase modulation of the display. The LCoS display was illuminated with $\lambda_G = 514$ nm.

Calibration curves were obtained for different configuration conditions of the SLM display in order to optimize the configuration parameters to achieve a variety of phase modulation ranges. The Holoeye modulator is electrically addressed by means of a standard digital video interface (DVI) connector and works with two bit-plane configurations (5-5 and 18-6), which allow mapping the 256 video grey levels to either 192 or 1216 index levels of the look-up table (LUT), respectively, loaded in the hardware of the device. Even though the number of addressable values for the 5-5 bit-plane profile is lower, this configuration is intended to reduce the flickering effects of the display. For this reason, the 5-5 bit-plane scheme was the chosen configuration for the calibration of the device, except for the case of the maximum 8π phase modulation, which was only achieved by the 18-6 bit-plane profile. Initially, the LUT (also named gamma curve) provided by the SLM manufacturer ("default gamma", curve with cross markers in Figure 3b) was used to control the display. Two electronic potentiometers were properly adjusted to obtain a given device response. By modifying the varying grey level among the 256 grey values, the phase modulation response was measured ("nonlinear ph. mod.", line with cross markers in Figure 3c). From this measurement, a new LUT was established ("modified gamma", line with circles in Figure 3b) in order to achieve a linear phase modulation ("linear ph. mod.", line with circles in Figure 3c). Figure 3 depicts the modified LUT and the corresponding linear performance for up to 8π phase modulation range for an illuminating wavelength of $\lambda_G = 514$ nm. In this condition, we measured the response of the display for another wavelength in the green region of the spectrum, $\lambda_G = 520$ nm (Figure 4). Note that the SLM phase modulation response with the new calibration LUT (or modified gamma curve) does not vary remarkably when a different, but close, illuminating wavelength is used. Thus, we assume that the phase modulation response is resistant to slight variations in the illuminating wavelength with respect to the one used to determine the new gamma curve.

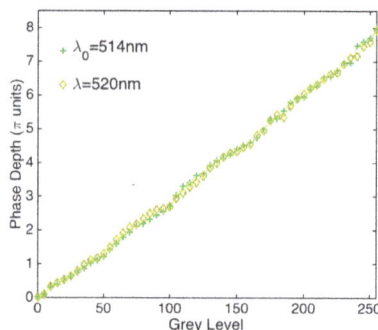

Figure 4. Phase modulation of the LCoS for two green wavelengths $\lambda_0 = 514$ nm (used in the characterization) and $\lambda = 520$ nm when the SLM display operates with the LUT that linearizes the 8π phase modulation range for $\lambda_0 = 514$ nm (Figure 3b,c).

Next, we measured the light reflected by the SLM display when a uniform image of a varying grey level was depicted as a full screen on the display. Light coming from the flat mirror was blocked at this time. The CCD camera was replaced by a photodetector so that the detector cell received the light reflected from the SLM display and integrated it. Figure 5a sketches the setup, and Figure 5b shows the modulation detected on the reflected light for the three illuminating wavelengths used sequentially in this experiment. This effect is due to the Fabry–Perot interference on the round-trip propagation inside the liquid crystal layer as already described in Refs. [6,14]. The number of oscillations in each plot accounts for the corresponding phase modulation range that approximates 6π, 8π, and 9π for $\lambda_R = 647$ nm, $\lambda_G = 514$ nm, and $\lambda_B = 476$ nm, respectively. As shown in Figure 4b, the experimental results obtained with our LCoS-SLM are in very good agreement with those reported by Calero et al. for another LCoS-SLM device (Hamamatsu X10468-08) [6].

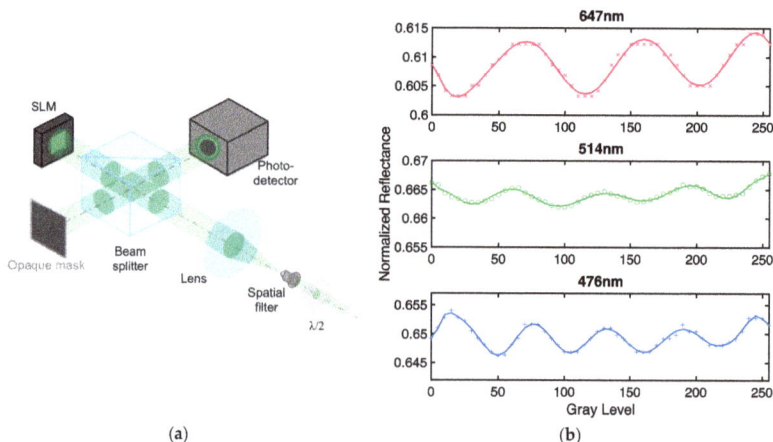

(a)

(b)

Figure 5. (a) Scheme of the experimental setup. (b) Reflectance modulation versus varying grey level addressed to the SLM.

We were able to obtain a variety of linear phase modulation ranges, from π up to 8π, by properly setting the configuration parameters of the SLM control driver. In particular, optimized calibration curves were obtained for π, 2π, 3π, 4π, 6π, and 8π phase modulation ranges. An analogous calibration procedure was followed in all cases. For those calibration tasks, the illuminating wavelength was set to 530 nm, which coincided with the maximum intensity of an available green LED source that was going to be used in subsequent experiments (display of Fresnel diffraction lenses). Once the calibration LUT was established, measurements for 647 and 476 nm, as representative of the available red and blue LED sources, were also taken.

From the plots of Figure 6, we remark that the phase modulation ranges were clearly linear as expected for the green source in all cases, and the phase modulation increases for shorter wavelengths because of the inverse dependence of the phase modulation with wavelength.

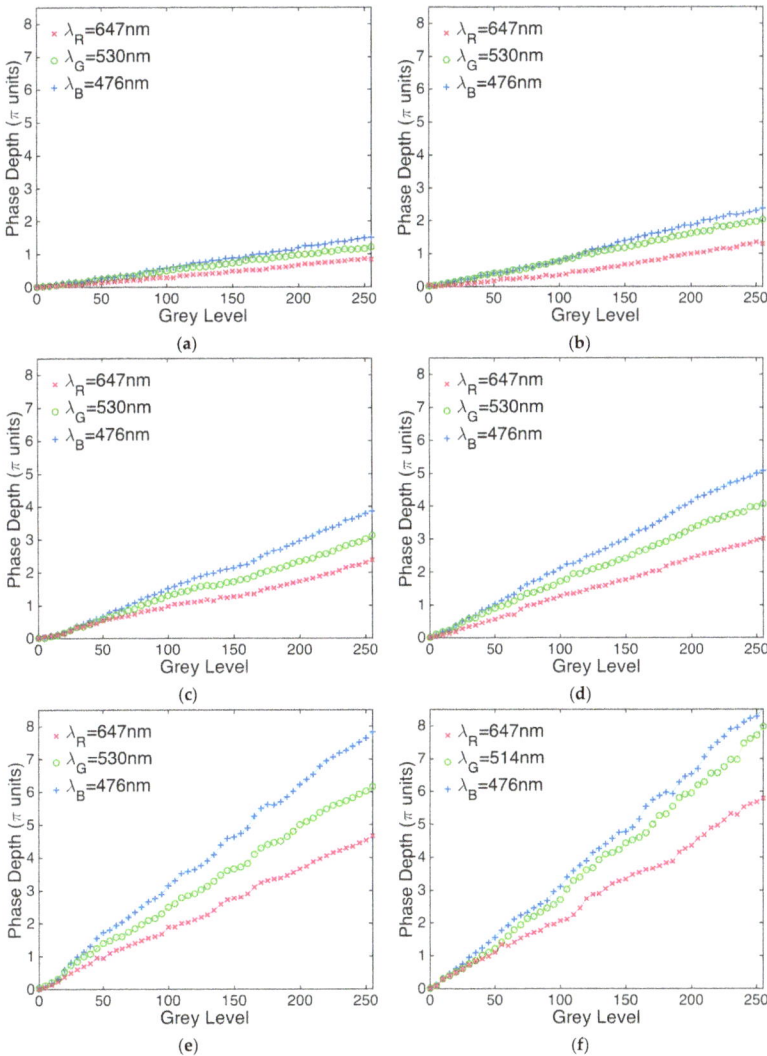

Figure 6. Phase modulation responses of the LCoS display for three different wavelengths (λ_R = 647 nm, λ_G = 530 nm, and λ_B = 476 nm). The calibration curves were optimized for λ_G = 530 nm and the phase modulation ranges: (**a**) 1π; (**b**) 2π; (**c**) 3π; (**d**) 4π; and (**e**) 6π while using the more stable SLM 5-5 bit-plane configuration. Note that the calibration curve for 8π phase modulation (**f**) was originally optimized for λ_G = 514 nm using the SLM 18-6 bit-plane configuration.

4. Overcoming the Nyquist Limit for a Lens with Low F-Number

Conventional modulo 2π diffractive lenses, which entail phase jumps of 2π for the design wavelength on either side of each zone boundary (Figure 1b), are designed to work efficiently in the first diffractive order [3,4]. One of the problems with the phase profile of a diffractive lens is that the phase function changes very rapidly at the outer part of the lens aperture. Thus, implementation is prone to undersampling and aliasing effects when the lens is displayed on a pixelated device (see, for instance, the bottom left corner of Figure 1b). As a rule of thumb, the range of focal lengths must meet the condition $f_N < f < 50 f_N$. The inferior limit is the Nyquist focal length [7,11], which is given by

$$f_N = \frac{N\Delta^2}{\lambda},\tag{4}$$

where Δ is the pixel pitch, and $N \times N$ is the size of the pixel array. Multiorder diffractive lenses with phase echelettes higher than modulo 2π (see Figure 1b) may overcome the limitations to implement first-order lenses. In that sense, new LCoS displays with large dynamic ranges in phase modulation offer advantages in comparison to older SLM technology.

If a first-order lens is displayed on a square area of $N = 1000$ pixels of the Holoeye LCoS SLM ($\Delta = 8$ microns pixel pitch) and is illuminated with $\lambda = 530$ nm, the corresponding Nyquist focal length—as derived from Equation (4)—will result in $f_N = 120$ mm (or equivalently, an optical power of 8.3 diopters). Thus, undersampling problems will arise at the periphery of the aperture if one implements a lens of shorter focal length (or higher optical power) on the display operating at a phase modulation range modulo 2π. In consequence, aliasing effects and poor performance would result. Let us overcome this situation by designing multiorder diffractive lenses.

Let us suppose we want to implement a diffractive lens with $f_0 = 80$ mm focal length (or, equivalently, 12.5 diopters). Note that $f_0 < f_N$ for the Holoeye LCoS-SLM. A possible solution would be to design a second-order lens of $f = 160$ mm that would obtain maximum efficiency on its second diffraction order located at $f/2 = f_0 = 80$ mm. Similarly, a third-order lens of $f = 240$ mm would concentrate the vast majority of the energy on its third diffraction order at $f/3 = f_0 = 80$ mm. Both multiorder diffraction lenses would have larger phase jumps at their edges in comparison to the conventional modulo 2π lens, and, in addition, both focal lengths (160 and 240 mm) would be above the Nyquist limit. Under these circumstances, the performances of second- and third-order lenses can be superior in comparison to conventional diffractive lenses.

We experimentally compared the performance of both multiorder lenses with their equivalent modulo 2π lens. We first considered a conventional lens with $f_0 = 80$ mm focal length for a design wavelength $\lambda_0 = 530$ nm. We displayed it on the LCoS display with the linear 2π calibration (Figure 6b) obtained for such wavelength. Note that the focal length of this lens is shorter than the focal length corresponding to the Nyquist limit ($f_N = 120$ mm), so that undersampling of the lens function will appear at the periphery of the lens. To avoid aliasing, we proposed a multiorder lens with larger focal length ($f > f_N$). We compared the performance of such a conventional lens with two different multiorder diffraction lenses: a second-order lens with $f = 160$ mm and a third-order lens with $f = 240$ mm computed for the design wavelength $\lambda_0 = \lambda_G = 530$ nm. The second-order diffraction lens was displayed on the LCoS by using the calibration LUT for a linear 4π phase variation (Figure 6d), whereas we implemented the third-order diffraction lens by using the linear 6π phase variation curve (Figure 6e) for the same design wavelength $\lambda_0 = \lambda_G = 530$ nm. Theoretically, all three lenses would have a common focalization plane at $f_0 = 80$ mm with 100% efficiency on its corresponding order of diffraction (first-order for the conventional modulo 2π lens, second-order for the modulo 4π lens, and third-order for the modulo 6π lens).

The setup sketched in Figure 7 was used in this experiment. It is similar to the Michelson interferometer used for characterization (Figure 2) but with some variations. In this experiment, a green LED source ($\lambda_G = 530$ nm) and a 100 micron pinhole were used to illuminate the setup, and a polarizer served to select the correct polarization plane of the light beam incident onto the SLM display.

Only the SLM arm was operative since light coming from the mirror was blocked with an opaque mask. And, finally, a microscope objective was attached to the CCD camera in order to acquire the focalization spots at the focal planes with appropriate magnification. This configuration, already used for device characterization, assures good control of the phase modulation for the on-axis operating lenses of our experiment (in the case of off-axis illumination with incident angle deviations within ±10 degrees, no substantial changes in the modulation properties would be expected either [12]). The three aforementioned diffractive lenses were sequentially displayed on the SLM screen, and the obtained experimental results are summarized in Figure 8. Figure 8a depicts the profile of the focalization spots located at $f_0 = 80$ mm for the three analyzed lenses. Profiles are cropped so that only the central area of the output zoomed plane is plotted.

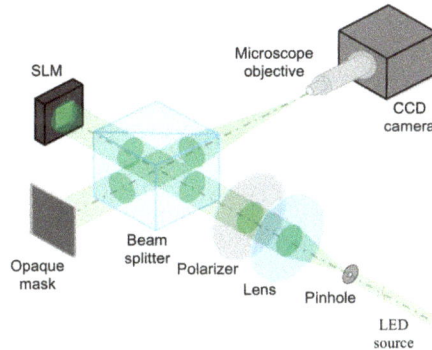

Figure 7. Setup for diffractive lens evaluation. Illumination is done with an LED source and a 100 micron pinhole. A polarizer selects the appropriate incidence on the SLM display. A microscope objective is attached to the charged coupled device (CCD) camera to capture the focal planes. Light coming from the mirror was blocked with an opaque mask.

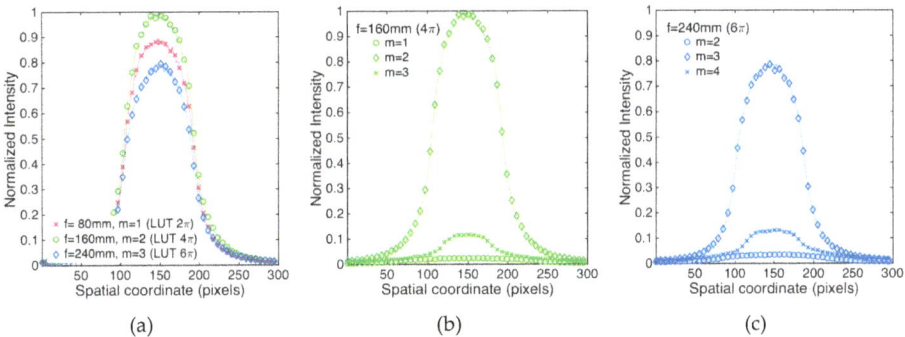

Figure 8. (**a**) Coincident focalization planes at $f_0 = 80$ mm for first-, second-, and third-order diffraction lenses. Calibrations curves for the linear 2π (Figure 6b), 4π (Figure 6d), and 6π (Figure 6e) responses were used, respectively. Different diffraction orders, at different locations, for (**b**) the second-order diffraction lens with $f = 160$ mm and (**c**) the third-order diffraction lens with $f = 240$ mm.

Intensity profiles plotted in Figure 8 were normalized to the maximum intensity peak obtained for the modulo 4π lens. As expected, the second-order lens produced a more intense focalization spot than the conventional modulo 2π lens, since the latter was below the Nyquist limit to reproduce the smallest details at the edges of the lens, whereas the modulo 4π lens avoided the aliasing effect. According to the graph, the focal spot of the modulo 2π lens was affected by an intensity reduction of about 10%

with respect to the modulo 4π lens. In the same Figure 8a, the intensity profile for the modulo 6π lens showed that this diffractive lens obtained the lowest focalization peak. It experienced a reduction around 20% of the maximum intensity. A possible reason for the latter result can be the fact that the LCoS display has a maximum of 256 grey levels to address the diffractive optics. The same number of grey levels is used to reproduce a phase range of 4π and 6π, so that the second-order lens can be implemented more precisely, in particular, at the central part of the lens aperture compared to the coarser reproduction of the third-order lens. Thus, the third-order lens would have larger quantization error than the second-order lens.

Figure 8b,c show the profiles of diffraction orders other than the one with a maximum efficiency for both the second-order lens and the third-order lens, respectively. In both cases, there was a small amount of light going to these orders, and both multiorder lenses had similar behaviors.

A second experiment was carried out to show the importance of using the optimized characterization curve in each case and to test the effect of the quantization error on the implementation of the lenses. The three different lenses were now displayed on the LCoS SLM by using a single calibration curve: the one corresponding to a linear 6π phase variation (Figure 6e). In this situation, only the full 256 grey level range was available to encode the 6π phase modulation of the third-order diffraction lens, whereas only parts of it were effectively used to encode the 4π phase modulation of the second-order diffraction lens (166/256 grey levels) or the 2π phase modulation of the first-order diffraction lens (just 66/256 grey levels). Figure 9 plots the profiles of the corresponding diffraction orders. They are all normalized to the maximum intensity peak shown in Figure 8a for the sake of comparison.

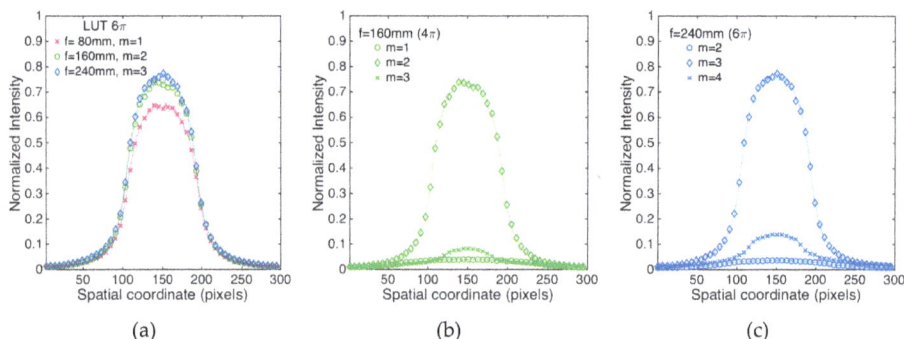

Figure 9. (**a**) Coincident focalization planes at f_0 = 80 mm for first-, second-, and third-order diffraction lenses when the 6π calibration curve (Figure 6e) is used in all the cases. Different diffraction orders, at different locations, for (**b**) the second-order diffraction lens with f = 160 mm and (**c**) the third-order diffraction lens with f = 240 mm.

In the analyzed situation, the most intense focalization peak was obtained for the third-order lens because its profile—with greater rings corresponding to a longer focal length and bigger facets at the periphery—could be encoded with the full grey level range. A slightly less intense peak (less than 5% of intensity reduction) was obtained for the second-order lens, mainly because it had a reduced number of phase levels (166 grey levels) to encode the 4π phase modulation. The focalization peak obtained for the conventional 2π lens had the lowest intensity among the profiles depicted in Figure 9a. The intensity reduction was around 15% compared to the third-order diffractive lens with full grey level range. The poor performance of the first-order lens can mainly be due to the combination of aliasing at the boundaries of the lens and the quantization error produced for the even more limited number of grey levels available.

Figure 9b,c show the different diffraction orders obtained at different planes for both multiorder lenses. In this case, when both lenses were displayed on the LCoS screen with the same calibration curve, which corresponds to 6π phase modulation, they obtained very similar results.

The results of Figures 8a and 9a lead us to remark the importance of using an optimized SLM calibration curve to display DOEs. For instance, the performance of both the modulo 4π lens with only 166 grey levels and the modulo 2π lens displayed with only 66 grey levels (Figure 9a) experienced an intensity reduction of about 25% with respect to the same lenses implemented with the full grey level range (Figure 8a).

5. Experimental Achromatic Diffractive Lens Implemented in an LCoS Display

In this section, we show the potential of multiorder lenses displayed on an LCoS-SLM with a large phase dynamic range to design achromatic diffractive lenses. In the exemplary study of this section, we considered a diffractive lens of 150 mm focal length under polychromatic illumination. The described Holoeye LCoS display permits a maximum phase modulation range of 8π (Figures 4 and 6f) for a wavelength in the green region of the visible spectrum (514, 520 nm), attaining a larger phase modulation depth in the blue region (476 nm) and approaching 6π in the red region (647 nm). By using the response curves of Figure 6f, we designed a four-order diffractive lens for a design wavelength $\lambda_0 = 470$ nm and $f = 600$ mm focal length that operates with maximum efficiency in its fourth diffraction order at $f/m = 600/4 = 150$ mm. This four-order diffractive lens under blue illumination has a phase profile that behaves, under illumination with $\lambda_R = 625$ nm, as a third-order diffractive lens of $f = 450$ mm that shows its maximum diffraction efficiency at $f/m = 430/3 = 150$ mm from the lens plane (Equations (2) and (3)). In other words, such a phase profile achromatizes the multiorder diffractive lens with a common plane of maximum efficiency for both the blue and red wavelengths.

We compared the performance of this multiorder lens, designed to be ideally achromatic, with a conventional modulo 2π lens, designed for $\lambda_0 = 530$ nm with a focal length $f_0 = 150$ mm, when both were illuminated under polychromatic light. These lenses were implemented on the LCoS display using either the calibration curve for a linear 8π phase modulation (Figure 6f) (achromatic multiorder lens) or the calibration curve for a linear 2π phase modulation (Figure 6b) (conventional modulo 2π lens). The setup depicted in Figure 7 was used in this experiment. We sequentially illuminated both lenses with three different LED sources whose featured wavelengths covered the visible spectrum ($\lambda_B = 470$ nm, $\lambda_G = 530$ nm, and $\lambda_R = 625$ nm) and acquired the focal spots at the planes of higher diffraction efficiency. Distance from the LCoS screen was measured in each case.

Figure 10 shows the experimental results. Figure 10a shows the focalization planes obtained for the conventional modulo 2π diffractive lens. At the location of the design focal plane ($f_0 = 150$ mm) the focal spot for the green LED ($\lambda_G = 530$ nm) was obtained. When the first-order diffractive lens was illuminated with either the red or blue LEDs, the focal plane moved in opposite direction with respect to the green wavelength. Both red and blue focal plane positions are indicated in Figure 10a. These results evidence the remarkable chromatic aberration of a diffractive lens, for which focal planes for the three R, G, and B testing wavelengths are located some distance apart from each other. In our experiment, the focal planes of the extreme wavelengths of the visible spectrum (625 and 470 nm) are separated around 41 mm.

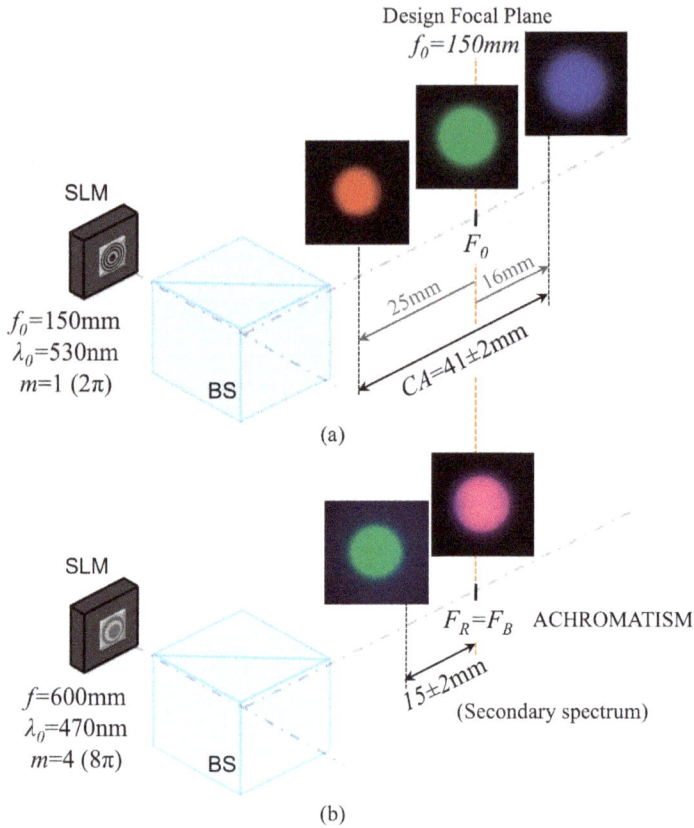

Figure 10. Experimental results for (a) Conventional modulo 2π diffractive lens for $\lambda_0 = 530$ nm and $f_0 = 150$ mm sequentially illuminated with three different wavelengths $\lambda_R = 625$ nm, $\lambda_G = 530$ nm, and $\lambda_B = 470$ nm. The focalization planes with the maximum diffraction efficiencies are shown at the location where they were acquired. (b) Analogous results for a fourth-order modulo 8π diffractive lens for $\lambda_0 = 470$ nm and $f = 600$ mm that worked in its fourth diffraction order located at $f/m = 600/4 = 150$ mm.

Figure 10b provides the experimental results for the achromatic multiorder lens. It consists of a fourth-order diffractive lens with a $f = 600$ mm focal length for a design wavelength of $\lambda_0 = \lambda_B = 470$ nm. The lens was sequentially illuminated by the three LED sources in the red, green, and blue regions of the visible spectrum. Focalization planes with the highest diffraction efficiencies were acquired, and their positions from the LCoS screen were measured. As a fourth-order diffractive lens ($m = 4$) for the blue light, the focalization plane with maximum efficiency for blue was located at $f/m = 600/4 = 150$ mm from the SLM display. Since this lens was, at the same time, a third-order diffractive lens for red light ($\lambda_R = 625$ nm), both bright blue and red spots appeared at the same position. The intensity spots captured by the CCD camera as well as their relative distances from the desired focal plane at 150 mm are shown in Figure 10b. These results demonstrate that the design multiorder lens has a resonant wavelength in the red region of the visible spectrum. Thus, for both wavelengths, $\lambda_0 = \lambda_B = 470$ nm and $\lambda_R = 625$ nm, the lens has a common focus at $f_0 = 150$ mm, which constitutes an achromatic lens. The light from the green LED, which is approximately in the middle of the spectrum, "sees" the diffractive lens with intermediate phase modulation steps, and the maximum intensity is obtained in a plane located 15 mm apart from the common red and blue foci. Thus, the secondary spectrum (residual chromatic aberration) for this multiorder lens drops to around one-third of the chromatic aberration measured for the conventional lens shown in Figure 10a.

6. Conclusions

A thorough characterization of a LCoS SLM display (Holoeye-Pluto-BB_HR) working on reflection has been carried out to obtain the optimized LUT curves for achieving different phase modulations, ranging from π up to 8π. Wavelengths from the green region of the visible spectrum (514, 520, and 530 nm) have been used to measure the various phase variation ranges and to compensate for some nonlinearities of the SLM response. From these initial measurements, the corresponding LUT curves were determined so that π, 2π, 3π, 4π, 5π, 6π, and 8π phase ranges were linearly reproduced by the SLM device. By using these phase modulation ranges, a number of multiorder diffractive lenses were displayed on the SLM device, and their performances were compared experimentally with the results of a conventional first-order diffractive lens. Second- and third-order diffractive lenses have proved to be very useful to overcome the undersampling problems raised when encoding a lens of relatively low f-number or optical power exceeding the Nyquist condition.

We have also shown the benefits of determining optimized LUT curves for each phase modulation range, since DOE performance exhibits better optical quality when the appropriate LUT with a full grey level range is used. According to the results obtained for high-order diffractive lenses, it is also preferable to have a sufficient range of available grey levels to smoothly display the phase function.

The so-obtained optimized LUT curves were used to determine the LCoS-SLM response for extreme wavelengths of the visible spectrum in the blue and the red regions. This information permitted the design of an achromatic, multiorder lens with a residual chromatic aberration reduced to one-third of the chromatic aberration of a conventional first-order diffractive lens. A fourth-order lens design for a blue wavelength ($\lambda_0 = 470$ nm) has a common focalization spot when it operates as a third-order diffractive lens for the red illumination wavelength $\lambda_R = 625$ nm, as demonstrated experimentally by our results.

Author Contributions: Conceptualization, methodology, validation, formal analysis, investigation, resources and writing—review and editing, M.S.M. and E.P.-C.; software, data curation and writing—original draft preparation, E.P.-C.; project administration and funding acquisition, M.S.M.

Funding: This research was funded by the Spanish Ministerio de Economía y Competividad and FEDER Funds (ref. DPI2016-76019-R).

Acknowledgments: Authors thank José Valverde and Joan Goset for their contribution in the experimental LCoS calibration procedure.

Conflicts of Interest: The authors declare no conflict of interest.

References

1. Sweeney, D.W.; Sommargren, G.E. Harmonic diffractive lenses. *Appl. Opt.* **1995**, *34*, 2469–2475. [CrossRef] [PubMed]
2. Faklis, D.; Morris, G.M. Spectral properties of multiorder diffractive lenses. *Appl. Opt.* **1995**, *34*, 2462–2468. [CrossRef] [PubMed]
3. Turunen, J.; Wyrowski, F. *Diffractive Optics for Industrial and Commercial Applications*, 1st ed.; Akademie Verlag (Wiley-Vch.): Berlin, Germany, 1997.
4. O'Shea, D.C.; Suleski, T.J.; Kathman, A.D.; Prather, D.W. *Diffractive Optics Design, Fabrication and Test*, 2nd ed.; Tutorial Texts in Optical Engineering; SPIE Press: Bellingham, WA, USA, 2015; Volume TT62.
5. Albero, J.; García-Martínez, P.; Martínez, J.L.; Moreno, I. Second order diffractive optical elements in a spatial light modulator with large phase dynamic range. *Opt. Lasers Eng.* **2013**, *51*, 111–115. [CrossRef]
6. Calero, V.; García-Martínez, P.; Albero, J.; Sánchez-López, M.M.; Moreno, I. Liquid crystal spatial light modulator with very large phase modulation operating in high harmonic orders. *Opt. Lett.* **2013**, *38*, 4663–4666. [CrossRef] [PubMed]

7.   Cottrell, D.M.; Davis, J.A.; Hedman, T.R.; Lilly, R.A. Multiple imaging phase-encoded optical elements written as programmable spatial light modulators. *Appl. Opt.* **1990**, *29*, 2505–2509. [CrossRef] [PubMed]

8.   Moreno, I.; Iemmi, C.; Márquez, A.; Campos, J.; Yzuel, M.J. Modulation light efficiency of diffractive lenses displayed in a restricted phase-mostly modulation display. *Appl. Opt.* **2004**, *43*, 6278–6284. [CrossRef] [PubMed]

9.   Arrizón, V.; Carreón, E.; González, L.A. Self-apodization of low-resolution pixelated lenses. *Appl. Opt.* **1999**, *38*, 5073–5077. [CrossRef] [PubMed]

10.   Yzuel, M.J.; Campos, J.; Márquez, A.; Escalera, J.C.; Davis, J.A.; Iemmi, C.; Ledesma, S. Inherent apodization of lenses encoded on liquid-crystal spatial light modulators. *Appl. Opt.* **2000**, *39*, 6034–6039. [CrossRef] [PubMed]

11.   Millán, M.S.; Otón, J.; Pérez-Cabré, E. Chromatic compensation of programmable Fresnel lenses. *Opt. Express* **2006**, *14*, 6226–6242. [CrossRef] [PubMed]

12.   Lizana, A.; Martin, N.; Estapé, M.; Fernandez, E.; Moreno, I.; Marquez, A.; Iemmi, C.; Campos, J.; Yzuel, M.J. Influence of the incident angle in the performance of Liquid Crystal on Silicon displays. *Opt. Express* **2009**, *17*, 8491–8505. [CrossRef] [PubMed]

13.   Otón, J.; Ambs, P.; Millán, M.S.; Pérez-Cabré, E. Multipoint phase calibration for improved compensation of inherent wavefront distortion in parallel aligned liquid crystal on silicon displays. *Appl. Opt.* **2007**, *46*, 5667–5679. [CrossRef] [PubMed]

14.   Davis, J.A.; Tsai, P.; Cottrell, D.M.; Sonehara, T.; Amako, J. Transmission variations in liquid crystal spatial light modulators caused by interference and diffraction effects. *Opt. Eng.* **1999**, *38*, 1051–1057. [CrossRef]

MDPI

St. Alban-Anlage 66

4052 Basel

Switzerland

Tel. +41 61 683 77 34

Fax +41 61 302 89 18

www.mdpi.com

Applied Sciences Editorial Office

E-mail: applsci@mdpi.com

www.mdpi.com/journal/applsci

www.ingramcontent.com/pod-product-compliance
Lightning Source LLC
Chambersburg PA
CBHW041215220326
41597CB00033BA/5978